MATHEMATICAL PROBLEM SOLVING:
ISSUES IN RESEARCH

Frank K. Lester and Joe Garofalo, editors

THE FRANKLIN INSTITUTE PRESS℠

© 1982 by THE FRANKLIN INSTITUTE

Printed in the United States of America

Published by THE FRANKLIN INSTITUTE PRESS^{SM}
Philadelphia, Pennsylvania

Current printing (last digit):
5 4 3 2 1

Library of Congress Cataloging in Publication Data

Mathematical Problem Solving.

Revisions of paper prepared for a conference held at Indiana University in May 1981, sponsored by the Spencer Foundation and the Indiana University School of Education.

Includes bibliographies.

1. Problem solving—Addresses, essays, lectures. 2. Mathematics—Study and teaching—Addresses, essays, lectures. 3. Mathematical research—Addresses, essays, lectures. I. Lester, Frank K. II. Garofalo, Joe. III. Spencer Foundation. IV. Indiana University, Bloomington. School of Education.

| QA63.M37 | 1982 | 510'.7 | 82-8609 |
| ISBN 0-89168-049-7 | | | AACR2 |

Contributors

Margaret Akerstrom
School of Education
Northwestern University
Evanston, IL

Diane J. Briars
Department of Psychology
Carnegie-Mellon University
Pittsburgh, PA

Joe Garofalo
Mathematics Education Department
Indiana University
Bloomington, IN

Gerald A. Goldin
Department of Mathematical Sciences
Northern Illinois University
DeKalb, IL

John F. LeBlanc
Mathematics Education Department
Indiana University
Bloomington, IN

Richard Lesh
School of Education
Northwestern University
Evanston, IL

Frank K. Lester
Mathematics Education Department
Indiana University
Bloomington, IN

Richard E. Mayer
Department of Psychology
University of California, Santa Barbara
Santa Barbara, CA

Richard J. Shumway
Faculty of Science and Mathematics Education
The Ohio State University
Columbus, OH

Harold L. Schoen
School of Education
University of Iowa
Iowa City, IA

Alan H. Schoenfeld
Department of Mathematics
University of Rochester
Rochester, NY

Edward A. Silver
Department of Mathematical Sciences
San Diego State University
San Diego, CA

Acknowledgments

We are grateful to several people for their help and suggestions in the preparation of this book. We wish to thank the editorial staff of The Franklin Institute Press for their assistance and a special note of thanks is due to Julia Hough, who was an enthusiastic supporter of this project from the beginning. We are grateful to Professor David C. Johnson and Dr. Kathleen Hart of the Centre for Science Education at Chelsea College, University of London. They provided invaluable suggestions and moral support to the first editor while he was preparing a working paper at the Centre during a sabbatical leave. This paper served as a key stimulus for the conference, "Issues and Directions in Mathematical Problem Solving Research," and subsequently this book.

We wish to acknowledge the Spencer Foundation and the Indiana University School of Education for their sponsorship of the above-mentioned conference and for their support for the work leading up to the conference.

Finally, we have enjoyed unusually good secretarial help and we are indebted to Jill Nicholas who typed the editors' contributions and assisted us in editing each chapter.

Contents

Introduction

The high level of interest in problem solving and thinking among educators and psychologists has been accompanied by a corresponding level of concern for developing a stable and useful body of knowledge about the phenomena associated with these highly complex areas of human behavior. There is no question that our understanding of problem-solving behavior has increased greatly in recent years. However, for various reasons, some of which are pointed out in Frank Lester's paper in this book, the research on the whole has been rather unsystematic and has lacked clarity of purpose and focus.

It has been said that a problem well formulated is half solved. We believe that a clearer formulation of the issues will be a big first step toward making the study of problem solving by math educators more systematic.

The purpose of this volume is to contribute to this clear formulation (of the key issues in mathematical problem-solving research) by presenting ideas of several individuals who are actively engaged in research in the area. With the exception of the contribution by Dick Shumway, the chapters represent revisions of papers prepared for a conference on "Issues and Directions in Mathematical Problem Solving" held at Indiana University in May, 1981. The conference was organized in response to an apparent interest within the mathematics education community in the establishment of better lines of communication about problem-solving research. Several individuals later suggested that the papers should be made available to an audience much wider than the conference participants. This book is the product of those suggestions.

All of the authors offer opinions as to where current and future research emphases should be placed and/or how such research should be conducted. Many of them argue convincingly that mathematics educators interested in problem solving need to draw upon the work being done by psychologists and still others address the problems inherent in the measurement of problem-solving outcomes.

The first paper, written by Rich Mayer, sees the poor performance of students on the national mathematics assessments of educational progress (NAEP) as a compelling reason for providing more effective mathematics instruction. Mayer, a cognitive psychologist who has been especially interested in "meaningful" learning in mathematics, states that psychologists have been challenged to create useful theories of learning because of this need for better mathematics instruction. In particular, he sees a need for establishing basic principles of learning, memory, and cognition that are relevant to tasks in mathematics. In other words, he is calling for a psychology of mathematics learning in general, and mathematical problem solving in particular. Mayer believes that cognitive psychology has already made contributions to such a theory of mathematical problem solving, one contribution being the distinction between two stages in problem solving: representation and solution. He also believes that cognitive psychology has identified types of knowledge that may be relevant to each of those two stages. Problem representation depends upon linguistic, factual, and schema knowledge, while problem solution depends on algorithmic and strategic knowledge. Mayer hopes for a partnership between math educators and psychologists, with the final product being a "fully ripened" psychology of problem solving.

In the next paper, Ed Silver argues that the role of the content and organization of a problem solver's knowledge base must be given substantial research attention. Silver, a math educator, is concerned that very little attention has been given to questions associated with knowledge organization and accessibility. His discussions of how schemata, cognitive elaboration, and metacognition influence knowledge organization draw upon the research of psychologists and are relevant to the two stages of problem solving mentioned by Mayer, primarily the representation stage.

In contrast to Silver, Alan Schoenfeld argues in the next paper that having a well-organized body of knowledge is not the most important ingredient in successful problem solving. Rather, he views the ability to use available resources strategically to be a more important factor. Schoenfeld emphasizes the solution stage of the two-stage model mentioned by Mayer.

Schoenfeld criticizes math educators for ignoring the work of psychologists and stresses the need for researchers to become familiar with it. He adds that caution is needed in interpreting such research since he believes that many psychological models of problem solving have confused proficiency with expertise. He also characterizes math educators as being "incestuous" for using the same limited set of problems in research and calls for the development of new problems for research purposes.

Like Schoenfeld, Diane Briars maintains that the work of psychologists has much to offer to the mathematics educator interested in problem solving. Briars, a mathematics educator who has been working closely with information-processing psychologists for the past two years, asks: "What new ideas and perspectives does information-processing psychology bring to the study of mathematics learning and problem solving?" In order to examine this question, she presents the fundamental concepts underlying an information-processing approach and discusses some recent advances in problem-solving research using this approach. She suggests that the most significant implications of this body of work are the identification of variables and constructs (e.g., limited short-term memory, knowledge organization, and distinctions between procedural and declarative knowledge) that are worth considering in future mathematical problem-solving research and the development of explicit models of behavior.

Like the authors before him, Frank Lester suggests that math educators would do well to look at cognitive psychology for direction and perspective and calls for the establishment of better channels of communication between math educators and psychologists. Lester poses seven questions as being among the most important for mathematical problem-solving research and implies that cognitive psychologists' research provides clues to the answers to at least three of the questions: *1.* What is the role of understanding in problem solving? *2.* To what extent does transfer of learning occur in problem solving? *3.* What are the most appropriate research methodologies to employ?

While the previous papers suggest that a math educator conducting problem-solving research would benefit from having the perspective of a psychologist, in the next paper Jerry Goldin illustrates how having the perspective of a physicist can benefit problem-solving research. As a physicist, he recognizes that the specific way a variable is operationalized in an investigation can have profound effects on results, interpretability, and generalizability. More specifically, he is concerned with the difficulties associated with measuring problem-solving outcomes. He insists that any definition of problem solving must include a description of how it is to be measured. He also argues that the scientific study of problem solving necessitates a deep understanding of the tasks used in the research. An especially valuable feature of Goldin's chapter is his discussion of the uses and limitations of the various kinds of measurements (scoring systems) that have been made in recent problem-solving investigations.

The measurement of problem solving outcomes is also one of the focal points addressed by Hal Schoen in the next paper. Schoen's paper is a set of reflections about measuring problem-solving performance by a math educator recently involved in developing a paper-and-pencil, group-administered mathematical problem-solving inventory. In particular, he raises several questions that researchers should consider as they devise measurement procedures. In addition to raising measurement questions, he suggests an application of certain information-processing tenets to research in teaching problem solving. It appears that Schoen is making a plea for problem-solving researchers to look at relevant teaching strategies research, a plea also made by Lester in a different volume (in press).

Following Schoen's appeal for problem-solving researchers to look at problem solving from the teacher's point of view is a paper by John LeBlanc, which presents a model for preparing elementary teachers to provide sound problem-solving instruction. This model evolved over several years of experience as an elementary mathematics teacher educator. While LeBlanc's paper addresses research questions only tangentially, it presents a point of view about preparing teachers to teach problem solving that should generate a number of researchable questions. It is evident from reading LeBlanc's ideas that he thinks teacher training in problem solving should be a top priority research area.

Although LeBlanc feels that problem-solving research has largely ignored teacher training, Dick Lesh and Margaret Akerstrom feel that most of the existing research is misdirected. One of their major premises is that the types of understandings, skills, and processes usually studied in problem-solving research are not necessarily the same as those needed to solve "real world" problems. They believe that the results of most problem-solving research, which employs typical textbook problems, has little or no relevance for solving problems involving realistic applications of mathematics. In their paper, Lesh and Akerstrom describe the activities and aims of two research projects concerned with "applied" problem solving — Applied Problem Solving Project and Rational Numbers Project.

The final paper, written by Dick Shumway, was solicited in order to include the ideas about problem-solving research of a mathematics educator with some "distance" from it. As a mathematics educator primarily interested in concept learning, he offers some personal biases about the nature of problem-solving research and suggests that mathematics educators would do well to give more attention to deciding what problem solving is and how it relates to other types of learning, concept learning in particular.

This book is intended for three types of researchers:

- Mathematics educators who desire direction regarding issues in problem-solving research;
- Psychologists involved in mathematical problem-solving research who are interested in the views of mathematics educators concerning problem-solving research issues;
- Educators in fields other than mathematics who also regard problem solving as an important, albeit complex, form of learning.

A final comment about the nature of this book is in order before concluding this introduction. Developments in problem-solving research happen quickly and the psychological literature is growing rapidly. A quick comparison of the papers written by Mayer and Lester points this out. No fewer than six of the main references discussed by Mayer were unavailable to Lester even though Lester's paper was written only seven or eight months prior to Mayer's. We have made a special effort to prepare this compilation of papers as quickly as possible while the ideas contained in them are still "hot" and "newsworthy." In doing this we have accepted the risk of sacrificing rigor and completeness for expediency. Nevertheless, we believe that the points of view and discussions proffered in these papers will prove to be at the same time provocative and valuable as a source of information about recent activities in mathematical problem-solving research.

Frank K. Lester
Joe Garofalo

The Psychology of Mathematical Problem Solving

Richard E. Mayer

The purpose of this paper is to describe what cognitive psychology has to say concerning instruction in mathematical problem solving. After an introduction that provides a definition, rationale, and historical perspective for a "psychology of mathematical problem solving," there follows an overview of the contributions of cognitive psychology, including techniques for analyzing knowledge.

Introduction

In a recently published book entitled *The Psychology of Mathematics for Instruction,* Resnick and Ford (1981, page 3) present the following discussion:

> As psychologists concerned specifically with mathematics, our goal is to ask the same questions that experimental and developmental psychologists ask about learning, thinking, and intelligence but to focus these questions with respect to a particular subject matter. What this means is that instead of asking ourselves a general question, "How is it that people think?" we ask ourselves, "How is it that people think about mathematics?" Instead of asking, "How do people's thought processes develop?" we ask, "How does understanding of mathematical concepts develop?"

Resnick and Ford also point out that a psychology of mathematical problem solving requires interdisciplinary efforts that combine "an understanding of the structure of the subject matter," with "knowledge of how people think in general . . . and how to study how people think."

Rationale

There is a widespread need to improve students' learning of mathematics. Mathematics is the foundation for many fields including science, engineering, business, and economics, and is vital to individuals' everyday survival as consumers. Our economic health and strength as a nation depend on our ability to develop new technology and to manage our businesses effectively. However, recent assessments of student achievement in mathematics are discouraging (California Assessment Program, 1979; Carpenter, et al., 1980). For example, over half of the twelfth graders in California public schools were unable to solve simple story problems such as the following (California Assessment Program, 1979): "An astronaut requires 2.2 pounds of oxygen per day while in space. How many pounds of oxygen are needed for a team of 3 astronauts for 5 days in space?" In the National Assessment of Educational

Progress (Carpenter et al., 1980), only 29 percent of a large national sample of 17-year-olds were able to solve the following problem: "Lemonade costs 95¢ for one 56 ounce bottle. At the school fair, Bob sold cups holding 8 ounces for 20¢ each. How much money did the school make on each bottle?"

The poor performance of students in story problems such as those given above points to the need for a better understanding of how to provide instruction in mathematics. The development of a psychology of mathematical problem solving can be the framework for improving learning of mathematics in schools.

Historical Perspective

In the past, psychological theories of learning and problem solving have either been on a very grand scale—such as Hull's general theory of the 1940s—or have been on a very narrow scale—such as recent research during the 1970s and 1980s in describing performance for specific tasks like the Tower of Hanoi. A psychology of mathematical problem solving is a compromise between a "general theory" covering all human cognition and a "special theory" that covers human cognition concerning a single task. It is a compromise because it allows for the development of general theory within a restricted well-defined domain.

The time seems right for the emergence of a psychology of mathematical problem solving. Previous attempts to build general theories proved unsuccessful and were abandoned by many psychologists, although much useful information was gathered along the way. Attempts to build special theories have been successful but often lack usefulness beyond the task that is analyzed. Furthermore, until recently such theories ignored the role of learning (Langley and Simon, 1981). Thus, a new research strategy is clearly needed. Recent research on problem-solving expertise has suggested that a theory of problem solving cannot be separated from the domain in which the problems are being solved (Tuma and Reif, 1980). In other words, a good theory requires an analysis of both the general problem-solving techniques and the domain-specific knowledge of the problem solver. It seems appropriate, then, that progress can be made when we focus on a domain—such as mathematics—rather than on *all* problem solving or on one individual puzzle.

The research strategy for a psychology of mathematics learning and problem solving has been spelled out recently in Resnick and Ford's *The Psychology of Mathematics for Instruction*. The book provides a detailed description of research in problem solving that is focused on mathematical tasks and points out that the psychology of mathematical problem solving is particularly attractive as a first "new frontier." This is because there is already a great deal of research concerning problem solving with mathematical tasks, because mathematics is an area that has practical importance, and because there exists a tradition of communication among math educators and psychologists. Thus, the psychology of mathematical problem solving has a head start. The next step, of course, is to take all the useful fragments at hand and develop an integrated theory of mathematical problem solving.

Contributions of Cognitive Psychology

Cognitive psychology is the study of how humans process information, and includes study of the acquisition, storage, and retrieval of knowledge. Humans are viewed generally as possessing *memory stores* such as short-term memory, working memory, long-term memory, and *memory processes* for transferring knowledge from one store to another or for manipulating information in memory (see Klatzky, 1980).

Cognitive psychology can offer a new approach to the development of a psychology of mathematical problem solving in three important ways.

First, cognitive psychology views learning as the acquisition of *knowledge,* rather than solely the acquisition of new *behaviors*. One of the major contributions of cognitive psychology concerns methods for breaking down knowledge into parts (Mayer, 1980, 1981a). Some of the types of knowledge that may be relevant for a psychology of mathematical problem solving are:

- Linguistic and factual knowledge—concerning how to encode sentences, such as grammar;
- Schema knowledge—concerning relations among problem types, such as "work problems" vs. "motion problems";
- Algorithmic knowledge—concerning how to perform well-defined procedures such as addition;
- Strategic knowledge—concerning how to approach problems.

The left portion of Table 1 lists these types of knowledge, and later sections of this paper explore the role of each type of knowledge in mathematical problem solving. There is some evidence that current instructional methods in mathematics emphasize acquisition of factual and algorithmic knowledge, while ignoring schema and strategic knowledge (Greeno, 1978; Anderson, et al., 1981).

Table 1. Some Types of Knowledge Needed in Mathematical Problem Solving

Type of Knowledge	Stage in Problem Solving Process	
	Representation (Understanding)	Solution (Searching)
Linguistic & Factual	x	
Schematic	x	
Algorithmic		x
Strategic		x

Second, cognitive psychology views problem solving as a series of mental operations that transform knowledge representations rather than viewing problem solving as a series of learned behaviors. A key concept is the problem space—a representation that contains the given state, the goal state, and all legal intervening states. To move from one state to another, a subject must apply an operator to the current state. One of the major contributions of cognitive psychology has been the distinction between two stages in problem solving (Mayer, 1977; Hayes, 1981; Wickelgren, 1974):

- Representation (understanding the problem)—The problem statement is translated into an internal mental representation that includes the given state, goal state, and allowable operators. The problem space can be built upon the subject's understanding of the problem.
- Solution (searching the problem space)—The problem solver attempts to search for a path through the problem space. The problem solver keeps applying operators to problem states until the goal is achieved.

The right portion of Table 1 lists these two stages in problem solving, and shows how the types of knowledge are used in each stage. Although the division between these two stages may not be as clean as suggested by the table, the distinction allows for some useful explanations of mathematical problem solving (Hayes and Simon, 1974). One interesting outcome of recent research is that difficulties in mathematical problem solving often result from inadequate representation of the problem (stage 1) rather than from faulty solution procedures (stage 2); however, most instruction stresses the solution process and subordinates knowledge of when to apply the procedures or how to represent problems (Simon, 1980).

Third, cognitive psychology suggests that instruction for problem solving should focus on *cognitive objectives*, i.e. descriptions of the to-be-learned knowledge structures, rather than focus on *behavioral objectives*, i.e. descriptions of the to-be-learned responses (Greeno, 1976). Another way to state this idea is to say that instruction should focus on *process*, i.e. how to solve problems, rather than *product*, i.e. the final answer (Bloom and Broder, 1950). The rationale for focusing on cognitive objectives or process is that two students may display the same performance (e.g., many errors in subtraction problems) but use entirely different procedures to achieve this behavior. For example, Brown and Burton (1978) found that students' performance in subtraction can be characterized by certain "bugs." Rather than saying that a student performed at a certain level of errors, Brown and Burton are able to provide the algorithm (including "bugs") that the child is using. This analysis may be useful for providing remediation. Similarly, Greeno (1976) has shown that students can learn to perform mathematical tasks, such as arithmetic with fractions, but use entirely different knowledge structures to generate correct answers. A fundamental distinction has been offered by the Gestalt psychologists (Wertheimer, 1959; Katona, 1942):

- Rote learning—The correct response or procedure is memorized without learner understanding.
- Meaningful learning—The procedure of solving problems is related to other knowledge, and understood.

Since it is much easier to define and measure rote learning, much emphasis has been placed on evaluating performance. However, there is much evidence that meaningful learning leads to superior transfer and long-term retention (Mayer, 1975).

The remainder of this section provides some examples of the role of linguistic and factual, schematic, algorithmic, and strategic knowledge in mathematical problem solving. In particular, examples will be drawn from research on how students learn to solve algebra story problems. Table 2 presents some typical story problems that were used in a recent study (Mayer, 1982). Let's consider what kinds of knowledge one would need in order to solve problems like these. The problems in Table 2 are, of course, not an exhaustive list of possible problems from the mathematics curriculum. However, they are presented as examples to illustrate the kinds of knowledge that students must acquire.

Table 2. Some Examples of Algebra Story Problems

River Problem
A river steamer travels 36 miles downstream in the same time that it travels 24 miles upstream. The steamer's engine drives in still water at a rate of 12 miles per hour more than the rate of the current. Find the rate of the current.

Freeway Problem
A truck leaves Los Angeles enroute to San Francisco at 1 p.m. A second truck leaves San Francisco at 2 p.m. enroute to Los Angeles along the same route. Assume the two cities are 465 miles apart and that the trucks meet at 6 p.m. If the second truck travels at 15 mph faster than the first truck, how fast does each truck go?

Frame Problem
The area occupied by an unframed rectangular picture is 64 square inches less than the area occupied by the picture mounted in a frame 2 inches wide. What are the dimensions of the picture if it is 4 inches longer than it is wide?

Mixture Problem
One vegetable oil contains 6% saturated fats and a second contains 26% saturated fats. In making a salad dressing how many ounces of the second may be added to 10 ounces of the first in order to make 16% saturated fats?

Linguistic and Factual Knowledge

In order to represent problems like those in Table 2—the first phase in the problem-solving process—a learner must know the rules of language (linguistic knowledge) and some basic facts about the world (factual knowledge). For example, in the river problem, a learner must know that "river steamer" and "it" and "steamer" all refer to the same object. In addition, a learner must be able to locate parts of speech such as variables or numbers or arithmetic operations. The learner must know what words mean; for example, in the freeway problem the learner must know that "enroute to Los Angeles" means that the truck goes from San Francisco to Los Angeles. Also, for the freeway problem the learner needs to know how to tell time, so that it can be determined that one truck travels for five hours while the other travels for four hours. Thus, linguistic and factual knowledge are needed in order to be able to translate a story into a set of equations.

A recent study (Mayer, 1982) shows that students have a great deal of trouble with linguistic constructions that involve relational comparisons. In the study, college freshmen were asked to read problems like those in Table 2 and to recall them later. Many of the errors in recall occurred when students changed a relational sentence, "The steamer's engine drives in still water at a rate of 12 mph more than the rate of the current," to a simpler non-relational sentence such as, "Its engines push the boat at 12 mph in still water." In another problem, many subjects changed the relational sentence, "The area occupied by an unframed rectangular picture is 64 square inches less than the area occupied by a picture mounted in a frame," to a non-relational sentence such as, "The area of an unframed picture is 64 inches." In the study, subjects made twice as many errors in remembering relational sentences as in remembering non-relational sentences. In a supplemental study, students were asked to make up story problems based on certain situations; in these problems, students tended to include almost *no* relational propositions.

Similarly, Greeno and his colleagues (Greeno, 1980; Heller and Greeno, 1978; Riley and Greeno, 1978) have tested whether certain kinds of propositions are more difficult to translate than others. For example, primary school children were quite proficient at repeating problems in which each sentence deals with one variable, such as a "cause/change" story: "Joe has 3 marbles. Then Tom gave him 5 more marbles. How many does Joe have now?" However, younger children made many errors when a sentence involved a relation between two variables such as in the "compare" story, "Joe has 3 marbles. Tom has 5 more marbles than Joe. How many marbles does Tom have?" Typically, students would repeat this problem as: "Joe has 3 marbles. Tom has 5 marbles. How many marbles does Tom have?" Apparently, children have more trouble comprehending and translating sentences that involve relational information.

In an earlier study, Loftus and Suppes (1972) located "structural variables" that affect the difficulty of story problems for sixth graders. For example, difficulty of a problem was increased if it was a different type from the previous one, if it required many arithmetic operations, and if the syntactic structure of the sentence was complex. It seems likely that difficulty was related to the specific structural properties of the sentences in the problem, although this idea was not directly tested. However, the findings cited above concerning the difficulties with relational information are consistent with Loftus and Suppes' findings that the hardest problem in their set was one containing a relational proposition: "Mary is twice as old as Betty was 2 years ago. Mary is 40 years old. How old is Betty?"

Clement, Lochhead, and Soloway (1979, 1980) have shown that difficulties in translating relational propositions are not limited to primary school children. College students were asked to write equations to represent propositions such as: "There are 6 times as many students as professors at the university." One-third of the students produced the wrong equation, with the most typical error being, "6S = P." However, when students were asked to translate relational statements like this one into a computer program, the error rate fell dramatically. Such results suggest that people have difficulty in interpreting relational propositions when they must use static formats such as equations or simple sentences.

Bobrow (1968) developed a computer program called STUDENT to translate certain types of word problems into equations, and then solve the equations. For the program to operate, it needs to be given information concerning how to recognize when two phrases refer to the same variable, how to locate variables and operators, and how to build simple sentences. In addition, the program needs additional factual information such as the number of feet in a mile, or the number of quarters in a dollar. Thus, linguistic and factual knowledge are needed in order to translate from words to equations—even when a computer is doing the translating.

Taken together, these results suggest that relational information is more difficult to represent mentally than other kinds of relevant information in a story. Relational propositions offer a severe road block to the subject's attempt to move from a story to an internal representation. This finding suggests that special attention should be paid to teaching children how to translate among relational propositions (in English), relational equations, and concrete manipulatives or pictures.

Schema Knowledge

Let's suppose that a learner possesses adequate amounts of linguistic and factual knowledge to translate story problems into equations. Is there anything else that the learner must know in order to accomplish the task of understanding the problem?

Recent research in our labs (Mayer, 1981b; 1982) has suggested that students' understanding of problems is influenced by their knowledge of problem types (or "schemas"). A problem schema is a general representation for a class of problems such as "motion" problems or "river current" problems. In a recent review of hundreds of problems from 10 basic algebra textbooks, Mayer (1981b) identified approximately 100 different problem types along with frequency norms for each, i.e. how often each type of problem appeared in textbooks. For example, there were 10 different types of current problems, 13 types of motion problems, etc., each type with its own distinctive solution procedure.

In a companion set of experiments (Mayer, 1982), subjects read story problems and later tried to recall them. Problems that were fairly common— i.e., types of problems that occurred frequently in textbooks—were recalled better than less common problems. The correlation between frequency and probability of correct recall reached between 66 percent and 85 percent in several analyses. In addition, errors in recall were often quite systematic, in which a low frequency problem was changed to a more common version of the problem. For example, the frame problem shown in Table 2 is a fairly unusual version of a frame problem; many subjects tend to recall it as a different version of a frame problem that is more often given in textbooks. In another experiment, subjects were asked to make up their own problems. In this experiment, almost all of the problems were high frequency problems, similar in form to problems in textbooks. These results show that students bring with them a knowledge of problem schemas; if a problem is unusual and does not correspond to a learned schema, students often translate that problem into a slightly different, more familiar problem.

Hayes and his colleagues (Hinsley, Hayes and Simon, 1977; Hayes, Waterman and Robinson, 1977; Robinson and Hayes, 1978) have obtained additional evidence concerning students' use of schemas for story problems. For example, in the Hinsley, Hayes, and Simon study, subjects were given problems to sort into groups. There was extensive agreement among the students concerning which problems went together. In all, they identified 18 distinct problem categories, including "distance-rate-time," "work," "motion," "triangle," "current," etc. When an ambiguous problem was presented to students, half interpreted it as a "triangle problem" and half as a "distance-rate-time problem." The two groups focused on entirely different information in the problem, and even misread facts in a way consistent with their categorization. For example, a subject who categorized his problem as a triangle one, misread "four minutes" as "four miles," assumed this was a leg of the triangle, and applied the Pythagorean theorem.

In other studies (Hayes, Waterman and Robinson, 1977; Robinson and Hayes, 1978), subjects were asked to judge which parts of a problem were relevant. Subjects tended to decide what category the problem belonged to, and then to make accurate judgments about which information was relevant. Apparently, what is remembered for a story problem is influenced by the subject's schema for the problem.

What happens when a student who lacks the appropriate schema tries to translate a problem? Paige and Simon (1966) investigated this question by asking students to solve "impossible" problems such as, "The number of quarters a man has is seven times the number of dimes he has. The value of the dimes exceeds the value of the quarters by $2.50. How many has he of each coin?" Some subjects translated the problem into equations, some subjects recognized the inconsistency, and some changed the problem to say that the value of the quarters exceeds the value of the dimes by $2.50, yielding the equation: $10X + 250 = 7(25X)$. Apparently, the first approach indicates that some subjects did not have a schema for the problem; the third approach suggests that some subjects tried to fit the problem within their past knowledge about similar problems.

Taken together, these results seem to suggest that in order to understand a problem, a student needs more than linguistic and factual knowledge. A student needs a way of structuring the information into a meaningful whole, e.g., by relating to a familiar problem type or schema. When a student is faced with a problem, the student may try to categorize the problem as belonging to a certain type. Once the student has determined the schema for the problem, the process of choosing relevant information and translating into equations can go on.

Algorithmic Knowledge

Once the student has understood the problem, that is, translated it into an internal representation, the next step is to solve the resulting equations. In order to solve equations, students need "algorithmic knowledge"—procedures that can be applied to equations. For example, in a recent set of studies (Mayer, Larkin and Kadane, 1980) students were asked to solve equations such as,

$$\frac{(8 + 3X)}{2} = 3X - 11$$

Some of the basic algorithms that the subject must be able to use in this problem are:

- Move—Adding, subtracting, multiplying or dividing both sides of the equation by the same number or variable, such as adding 11 to both sides or subtracting 3X from both sides.
- Compute—Combining two numbers or two variables that are both on the same side of the equation, such as changing 22 – 8 to 14 or changing 6X – 3X to 3X.

Results from a long series of experiments indicate that these algorithms are fairly automatic in college students. There were very few errors in applying these algorithms, and the time required for a "move" was about 1.5 seconds while the time required for a "compute" was about 1 second.

Lewis (1981) has found several interesting differences between experts and novices in applying algebraic operators, including the tendency of experts to use more complex procedures. In addition, several researchers have modeled algorithms used in arithmetic, and found that there are wide individual differences in the procedures that people use to generate answers in addition and subtraction (Brown and Burton, 1978; Groen and Parkman, 1972; Woods, Resnick, and Groen, 1975).

Algorithmic knowledge of the mechanical procedures in algebra and arithmetic are often heavily emphasized in mathematics instruction (Simon, 1980). Certainly, algorithmic knowledge is an important component of the skill required in solving problems. However, Simon argues that other forms of knowledge are sometimes not taught, and that algorithmic knowledge alone will not sustain creative problem solving. Students must also know "when" to apply operators as well as "how" to apply them.

Strategic Knowledge

Problem solving also requires knowledge of strategies. Students need techniques that will help them develop plans for solution (Polya, 1968). For example, Mayer, Larkin, and Kadane (1980) identified two distinct strategies for solving single-valued equations. "Isolate variable" involved trying to get all the X's on one side and all the numbers on the other side; "reduce expression" involved trying to combine variables or numbers so as to make the equation smaller. These two strategies lead to quite different patterns of performance. Similarly, Mayer and Greeno (1975; Mayer, 1978) found that subjects could use different strategies in solving three simultaneous equations. For example, some subjects worked forward from the givens to the goal while others worked backwards from the goal towards the givens (Wickelgren, 1974).

Recent work in equation solving by experts and novices indicates that different strategies are often used (Lewis, 1981; Larkin, 1981). Bundy (1975) has suggested several different strategies used in artificial intelligence for solving equations, and Matz (1981) has provided some empirical evidence concerning the reality of these strategies. Finally, Anderson, et al. (1981) have provided new evidence of the importance of "strategic knowledge" in geometry proofs. Again, there is very little direct instructional work on teaching effective problem-solving strategies, in spite of the critical importance of strategic knowledge in problem solving.

Conclusion

In a recent review of research on learning cognitive skills, Anderson (1981) noted that cognitive psychology has for the past 20 years "abandoned interest in learning." Thus, without a theory of learning, cognitive psychology has not had much to contribute to a theory of instruction. Happily, this trend is coming to an end. Pushed forward by the need to provide effective instruction in mathematics, cognitive psychology has been challenged to develop a useful theory of learning. My hope is that there will be a continued partnership between mathematics educators and cognitive psychologists, each challenging the other, with the final product being a fully ripened psychology of mathematical problem solving.

References

Anderson, J. R. (Ed.), *Cognitive Skills and Their Acquisition*. Hillsdale, NJ: Lawrence Erlbaum Associates Inc., 1981.

Anderson, J. R., Greeno, J. G. Kline, P. J., and Neves, D. M. "Acquisition of Problem Solving Skill." *In* J. R. Anderson (Ed.), *Cognitive Skills and Their Acquisition*. Hillsdale, NJ: Lawrence Erlbaum Associates Inc., 1981.

Bloom, B. S. & Broder, L. J. *Problem-solving Processes of College Students*. Chicago: University of Chicago Press, 1950.

Bobrow, D.G. "Natural Language Input for a Computer Problem-solving System." *In* M. Minsky (Ed.), *Semantic Information Processing*. Cambridge, MA: MIT Press, 1968.

Brown, J. S. and Burton R. R. "Diagnostic Models for Procedural Bugs in Basic Mathematical Skills." *Cognitive Science*, 1978, *2*, 155-192.

Bundy, A. "Analyzing Mathematical Proofs." Edinburgh: University of Edinburgh, Department of Artificial Intelligence, Research Report No. 2, 1975.

California Assessment Program. *Student Achievement in California Schools:* 1978-79 Annual Report. Sacramento, CA: State Department of Education, 1979.

Carpenter, T. P., Corbitt, M. K., Kepner, H. S., and Lindquist, M. M. "National Assessment: A Perspective of Mathematics Achievement in the United States." *In* R. Karplus (Ed.), *Proceedings of the Fourth International Conference for the Psychology of Mathematics Education*. Berkeley, California: International Group for the Psychology of Mathematics Education. Berkeley, California: International Group for the Psychology of Mathematics Education, 1980.

Clement, J., Lochhead, J. and Soloway, E. "Translating Between Symbol Systems: Isolating a Common Difficulty in Solving Algebra Word Problems." (COINS Technical Report 79-19) Amherst, MA: University of Massachusetts, Department of Computer and Information Sciences, March, 1979.

Clement, J., Lochhead, J., and Soloway, E. "Positive Effects of Computer Programming on Students' Understanding of Variables and Equations." Paper presented at National Conference of the Association for Computing Machinery, 1980.

Greeno, J. G. "Cognitive Objectives of Instruction: Theory of Knowledge for Solving Problems and Answering Questions." *In* D. Klahr (Ed.), *Cognition and Instruction*. Hillsdale, NJ: Lawrence Erlbaum Associates Inc., 1976.

Greeno, J. G. "A Study of Problem Solving." *In* R. Glaser (Ed.), *Advances in Instructional Psychology*. Hillsdale, NJ: Lawrence Erlbaum Associates Inc., 1978.

Greeno, J. G. "Some Examples of Cognitive Task Analysis With Instructional Implications." *In* R. E. Snow, P. Frederico, and W. E. Montague (Eds.), *Aptitude, Learning, and Instruction*. Hillsdale, NJ: Lawrence Erlbaum Associates Inc., 1980.

Groen, G. J. and Parkman, J. M. "A Chronometric Analysis of Simple Addition." *Psychological Review,* 1972, 79, 329-343.

Hayes, J. R. *The Complete Problem Solver.* Philadelphia: The Franklin Institute Press, 1981.

Hayes, J. R. and Simon, H. A. "Understanding Written Instructions." *In* L. W. Gregg (Ed.), *Knowledge and Cognition.* Hillsdale, NJ: Lawrence Erlbaum Associates Inc., 1974.

Hayes, J. R., Waterman, D. A. and Robinson, C. S. "Identifying Relevant Aspects of a Problem Text." *Cognitive Science,* 1977, *1,* 297-313.

Hinsley, D., Hayes, J. R. and Simon, H. A. "From Words in Equations." *In* P. Carpenter and M. Just (Eds.), *Cognitive Processes in Comprehension.* Hillsdale, NJ: Lawrence Erlbaum Associates Inc., 1977.

Heller, J. and Greeno, J. G. "Semantic Processing in Arithmetic Word Problem Solving." Paper presented at the Midwestern Psychological Association, 1978.

Katona, G. *Organizing and Memorizing.* New York: Columbia University Press, 1942.

Klatzky, R. *Human Memory* (2nd ed). San Francisco: W. H. Freeman & Co. Publishers, 1980.

Langley, P. and Simon, H. A. "The Central Role of Learning in Cognition." *In* J. R. Anderson (Ed.), *Cognitive Skills and Their Acquisition.* Hillsdale, NJ: Lawrence Erlbaum Associates Inc., 1981.

Larkin, J. "Enriching Formal Knowledge: A Model for Learning to Solve Textbook Physics Problems." *In* J. R. Anderson (Ed.), *Cognitive Skills and Their Acquisition.* Hillsdale, NJ: Lawrence Erlbaum Associates Inc., 1981.

Lewis, C. "Skill in Algebra." *In* J. R. Anderson (Ed.), *Cognitive Skills and Their Acquisition.* Hillsdale, NJ: Lawrence Erlbaum Associates Inc., 1981.

Loftus, E. F. and Suppes, P. "Structural Variables That Determine Problem-solving Difficulty in Computer-assisted Instruction." *Journal of Educational Psychology,* 1972, *63,* 531-542.

Matz, M. "Towards a Computational Theory of Algebraic Competence." *Journal of Mathematical Behavior,* 1980, *3,* 93-166.

Mayer, R. E. "Information Processing Variables in Learning to Solve Problems." *Review of Educational Research,* 1975, *45,* 525-541.

Mayer R. E. *Thinking and Problem Solving.* Glenview, IL: Scott, Foresman & Co., 1977.

Mayer, R. E. "Effects of Meaningfulness on the Representation of Knowledge and the Process of Inference for Mathematical Problem Solving." *In* R. Revlin and R. E. Mayer (Eds.), *Human Reasoning,* Washington: Winston-Wiley, 1978.

Mayer, R. E. "Begle Memorial Series on Research in Mathematics Education: Research on Memory and Cognition." *In Proceedings of the Fourth International Congress on Mathematics Education.* Berkeley, California: ICME, 1980.

Mayer, R. E. *The Promise of Cognitive Psychology.* San Francisco: W. H. Freeman & Co. Publishers, 1981 (a).

Mayer, R. E. "Frequency Norms and Structural Analysis of Algebra Story Problems into Families, Categories, and Templates." *Instructional Science,* 1981, *10,* 135-175 (b).

Mayer, R. E. "Memory for Algebra Story Problems." *Journal of Educational Psychology,* 1982, 74, in press.

Mayer, R. E. and Greeno, J. G. "Effects of Meaningfulness and Organization on Problem Solving and Computability Judgments." *Memory and Cognition,* 1975 *3,* 356-362.

Mayer, R. E., Larkin, J. H. and Kadane, J. "Analysis of the Skill of Solving Equations." Santa Barbara: Department of Psychology, Series in Learning and Cognition, Report No. 80-2, 1980.

Paige, J. M. and Simon, H. A. "Cognitive Processes in Solving Algebra Word Problems." *In* B. Kleinmuntz (Ed.), *Problem Solving: Research, Method and Theory.* John Wiley & Sons Inc., 1966.

Polya, G. *Mathematical Discovery.* New York: John Wiley & Sons Inc., 1968.

Resnick, L. B. and Ford, W. W. *The Psychology of Mathematics for Instruction.* Hillsdale, NJ: Lawrence Erlbaum Associates Inc., 1981.

Riley, M. S. and Greeno, J. G. "Importance of Semantic Structure in the Difficulty of Arithmetic Word Problems." Paper presented at the Midwestern Psychological Association, 1978.

Robinson, C. S. and Hayes, J. R. "Making Inferences About Relevance in Understanding Problems." *In* R. Revlin and R. E. Mayer (Eds.), *Human Reasoning.* Washington: Winston/Wiley, 1978.

Simon, H. A. "Problem Solving and Education." *In* D. T. Tuma and F. Reif (Eds.), *Problem Solving and Education: Issues in Teaching and Research.* Hillsdale, NJ: Lawrence Erlbaum Associates Inc., 1980.

Tuma, D. T. and Reif, F. *Problem Solving and Education: Issues in Teaching and Research.* Hillsdale, NJ: Lawrence Erlbaum Associates Inc., 1980.

Wertheimer, M. *Productive Thinking.* New York: Harper & Row, Publishers Inc., 1959.

Wickelgren, W. *How To Solve Problems.* San Francisco: W. H. Freeman & Co. Publishers, 1974.

Woods, S. S., Resnick, L. B., and Groen, G. J. "An Experimental Test of Five Process Models for Subtraction." *Journal of Educational Psychology,* 1975, *67,* 17-21.

Knowledge Organization and Mathematical Problem Solving*

Edward A. Silver

George Polya (1973) has written that a "well-stocked and well-organized body of knowledge is an asset to the problem solver. Good organization which renders the knowledge readily available may be even more important than the extent of the knowledge." Without ranking organization above extensiveness of knowledge, Herbert Simon (1980) has noted that "research on cognitive skills has taught us . . . that there is no such thing as expertness without knowledge—extensive and accessible knowledge" (p. 82).

Most researchers interested in mathematical problem solving would probably agree that the content and organization of a solver's "knowledge base" is a critical factor to consider in problem-solving research. In order to interpret correctly the problem-solving behaviors of an individual engaged in a problem solution episode, it is helpful to have considerable background information about his knowledge base. Moreover, to understand why a problem-solving instructional routine was or was not effective for a particular subject, it is necessary to know the nature of the knowledge structures into which that subject assimilated the instructional information. However, there is very little evidence of explicit attention to these issues in mathematical problem-solving literature. Perhaps because they are of such obvious importance, knowledge organization and accessibility issues related to previously learned mathematics content and previously solved problems have been ignored by most researchers. Even less attention has been given to the issues relating to the to-be-learned material or the to-be-solved problem in problem-solving instructional studies.

This paper discusses several possible influences on knowledge organization and accessibility. In particular, it deals with two cognitive phenomena (mathematical schemata and cognitive elaboration), and two metacognitive phenomena (strategy selection and belief systems), as they relate to knowledge organization for mathematical problem solving. This paper discusses briefly some of the available psychological theory and attempts to relate the available formulations to mathematics. The primary emphasis, however, is placed on presenting ways in which cognitive and metacognitive factors can influence the encoding and retrieval of information, and thereby influence mathematical problem-solving behavior.

---

*This paper was supported in part by National Science Foundation Grant No. SED 80-19328. Any opinions, conclusions, or recommendations expressed are those of the author and do not reflect the views of the National Science Foundation.

The Role of Schemata

The notion of a memory schema (a cluster of knowledge that describes the typical properties of the concept it represents) has proved quite useful in recent years in explaining many aspects of human knowledge organization and recall, especially in the area of prose text learning. In the past five years or so, a considerable amount of research has been generated on the influence and use of schemata. According to the usual discussions of schemata by psychologists (e.g., Thorndyke and Yekovich, 1980), a schema represents a prototypical abstraction of a complex and frequently encountered concept or phenomenon, and it is usually derived from past experience with numerous exemplars of this concept. Schemata have been shown to be associated with not only interpreting and encoding incoming information, but also recalling previously processed information (Thorndyke and Hayes-Roth, 1979). For example, it has been demonstrated that a single text passage often permits multiple interpretations based on different schemata possessed by the readers (e.g., Anderson, Reynolds, Schallert, and Goetz, 1977), and that schemata can influence the recall of information from memory (e.g., Mandler and Johnson, 1977). Furthermore, it has been argued that schemata can account for the inferences that are made in the face of incomplete information (e.g., Bransford, Barclay, and Franks, 1972).

The first explicit link between schema theory and mathematical problem solving was made by Hinsley, Hayes, and Simon (1977). In their series of studies, subjects were instructed to do the following:

- Categorize a series of standard algebra word problems into groups of related problems;
- Categorize a given problem after hearing only a portion of the problem text;
- Solve a set of nine algebra word problems. Three of these were standard problems and six were non-standard, that is, their cover stories and underlying structures did not match;
- Solve two "nonsense problems" that were constructed by taking standard algebra problems and replacing some of the content words with nonsense words;
- Solve the "small town" problem, a standard distance-rate-time problem to which some irrelevant information concerning the right triangular relation of three problem elements was added.

Hinsley et al. reported that subjects sorted the problems into standard categories, such as distance-rate-time or age problems, and that subjects were able to categorize a problem almost immediately, usually after hearing only the first few words. For example, after hearing the three words, "A river steamer...," one subject said, "It's going to be one of those river things with upstream, downstream, and still water. You are going to compare times upstream and downstream—or if the time is constant, it will be the distance." Furthermore, they reported that subjects tended to use problem categorizations to assist in retrieving useful solution information from long-term memory. Moreover, in the problems that were susceptible to multiple interpretation, subjects who categorized the problems as a distance-rate-time problem attended to different information in the problem statement than did subjects who categorized it as a Pythagorean triangle problem. In general, Hinsley et al. concluded that their subjects did have schemata for standard algebra problems and that the schemata influenced the encoding and retrieval of information during problem solving.

Mayer (Note 1) recently conducted a series of experiments in which subjects read a set of standard algebra word problems, and then were asked to recall each problem and to construct problems based on certain situations (e.g., "trains leaving stations"). He found that subjects recalled relevant information much better than irrelevant details, recalled high frequency problem forms (i.e., common standard problem types) better than low frequency forms, made recall errors that converted low frequency forms to high frequency forms, and constructed problems that matched standard textbook forms. He interpreted his findings as supporting the hypothesis that students possess schemata for standard algebra problems, and that the schemata guide the encoding and retrieval of problem information.

It is possible to develop schema-based interpretations of several common problem-solving phenomena. For example, the finding of Loftus and Suppes (1972) that a word problem in a sequence of problems is more difficult to solve if it is preceded by problems of a different type and Luchins' (1942) observations of *einstellung* or problem-solving set in the famous "water jar" problems can both be explained in terms of difficulties in shifting from one schema to another. Furthermore, there is a straightforward schema-based explanation of a student who has difficulty solving any problem that deviates from a standard textbook form.

Bob Davis and his colleagues (Davis, Jockusch, and McKnight, 1978; Davis and McKnight, 1979) have written extensively about schemata and the related notion of a "frame." They have demonstrated that these ideas may explain many aspects of algebraic task performance, including, but not limited to, the solution of algebra word problems. Jim Greeno and his colleagues (Heller and Greeno, Note 2; Riley and Greeno, Note 3) have identified schemata for elementary arithmetic word problems, and different formulations of schemata have been developed by Briars and Larkin (Note 4). As yet, however, we have few instantiations of schemata in other mathematical domains, especially sophisticated ones. Promising areas for schema research include theorem proving in geometry, mental calculation, and differentiation and integration in elementary calculus.

The next wave of research in this area must go beyond mere instantiations of mathematical schemata, beyond merely observing that persons tend to behave in stereotypic ways when presented with highly-standardized tasks. We need to know a great deal more about the development of schemata and how it may be positively influenced. On the basis of current schema research findings, it is premature to develop instructional implications for teaching problem solving. We need to be mindful of the Soviet research (e.g., Yaroschuk, 1957/1969) suggesting that the teaching of model problems and the requirement that students identify a problem's "type" before solving it may not always be useful to problem solvers, and may sometimes lead to the undesirable consequence of only attending to features of a problem's statement and not to a problem's mathematical structure. Since evidence shows that many students attend to superficial aspects of a problem statement (Silver, 1979), we need more information on the development of useful problem schemata before making instructional suggestions.

Valuable extensions of current theory could be obtained not only from studies of schema development but also from careful observation of individual differences in schema content or schema utilization. In particular, it may be productive to explore the differences between the problem schemata possessed by highly skilful problem solvers in a task domain and those possessed by less skilful solvers. In addition, it might be useful to study schema differences among problem solvers who are similarly proficient; differences between two "experts" in a task domain may be as important to study as the differences between "experts" and "novices" in that domain. For example, a person who would not sort standard algebra problems into groups of distance-rate-time, mixture, and river current problems, but instead put them together into one category—generalized rate problems— would have a very different set of problem schemata from those subjects studied by Mayer and by Hinsley et al. Such a person would have organized his or her algebra problem-solving experience in a very powerful manner.

It is likely that skilful problem solvers are able to use any of several organizational schemata for a given problem (i.e., the problem is multiply-coded in memory). What characteristics of a problem trigger the activation of one particular schema? In general, we need more information concerning the activities of schemata during problem comprehension and information encoding and retrieval.

We have seen that mathematical schemata can have important influences in knowledge organization and the processes of information encoding and retrieval. Schema theory has the potential to offer powerful explanations of mathematical problem solving on highly standardized tasks. Mathematical schemata may be a rich domain in which to investigate the structure and processes of routine problem solving. Nevertheless, it is unlikely that schemata will provide the basis for a complete theory of mathematical problem solving. Not all the problems that students of mathematics solve are routine and standard, and they are certainly not amenable to schema-based interpretations unless they have been frequently encountered. Thus, we need to look at other influences on knowledge organization that might explain how problem solvers integrate new information into memory and successfully confront nonroutine problems.

The Role of Elaboration

One way that new information is processed is through the use of cognitive elaboration. When using this learning strategy, the learner creates a symbolic construction that combines with the new information to make it more meaningful (Rohwer, 1970). Elaboration refers not only to the processes involved in relating new material to previous knowledge either directly or by analogy, but refers also to the creation of logical relationships among components of the material and drawing inferences or implications (Weinstein, Underwood, Wicker, and Cubberly, 1979). One explanation for the success of elaboration strategies is that they make the new information more meaningful by forming a relationship between the new, unfamiliar material and the old, already-learned information.

Rohwer (1976) has reviewed a large number of studies involving the use of pictures and imagery in learning from prose text. He suggested not only that elaboration during learning fosters long-term storage of the information but also that many individual and developmental differences may be explained by the tendency and ability of the individuals to elaborate. Reder (1980) has recently reached similar conclusions. Spiro (1977), in discussing the reconstructive aspects of human memory, has suggested that education focus on knowledge updating rather than compartmentalization, relating text to pre-existing cognitive structures rather than interrelating within text. This is consistent with Ausubel's (1968) theory of meaningful learning.

Further support for the importance of elaboration comes from a series of studies concerning the organization of information for meaningful learning conducted by Richard Mayer and his associates. This work (Mayer and Greeno, 1972; Mayer, 1976, 1977) demonstrated that elaborative organization of information during learning (forming connections with pre-existing cognitive structures or memory representations, i.e., "external connectedness") results in superior learning performance when compared with an approach emphasizing the interrelationships of concepts in the to-be-learned material ("internal connectedness") or when compared with an approach emphasizing organizational activity at the time of retrieval (Mayer, Note 5).

Although most research on elaboration has focused on learning from prose text, there are several ways in which elaboration theory might be applied to mathematical problem solving. We can probably use available theory on cognitive elaboration to design more effective instruction in mathematical problem solving. For example, by pointing out the ways in which a given problem is similar to or different from other problems a student has encountered in an instructional program, we increase the probability that the student not only will retrieve useful information from memory about related problems, but also will encode information about this problem solution episode in appropriate and useful ways for future reference. Furthermore, individual differences in elaborative behaviors may explain the success of some students in an instructional program that fails to help other students.

Bob Davis (Davis, Jockusch, and McKnight, 1978) has suggested that high-ability students often think about and rework problems they have just encountered in a test situation. Such behavior is likely to elaborate connections between the problems and previously learned material and leads to a more powerful organization of knowledge. The identification and careful description of other elaboration behaviors of highly capable students should increase our understanding of mathematics problem solving and mathematics learning in general.

Furthermore, it might be productive to consider some of Polya's heuristic suggestions as elaborative prompts. For example, the suggestions to "think of a related problem" and "look back at your solution" are elaborative in nature. In particular, "looking back" identifies aspects of a problem and its solution that connect that problem in memory to previously acquired knowledge. This in turn increases the probability that such connections occur, and further establishes knowledge in long-term memory that may be elaborated in later problem-solving encounters. It is well established that problem-solving instruction emphasizing "looking back" has had little success in inducing students to "look back" without teacher guidance. This vexing finding has led many to despair over the efficacy of "looking back" instruction. Perhaps we would reach a different conclusion if we viewed "looking back" as an elaborative prompt and looked for its effects not in terms of overt "looking back" behaviors by students solving subsequent problems, but rather in the ways in which students approach future problems. From an elaborative point of view, "looking back" at a problem today is preparation for one to "think of a related problem" tomorrow. By paying careful attention to elaboration theory, many of the instructional suggestions designed to help students use Polya's heuristic advice (e.g., Silver and Smith, 1980) could be refined into efficient instructional routines.

Thus far we have discussed only cognitive influences on knowledge organization. Although the use of an elaboration strategy is a cognitive action, the *decision* to do so is essentially metacognitive. It is increasingly clear that the behavior of a person engaged in complex problem solving is governed not only by the available cognitive resources (e.g., algorithms, heuristics, facts, schemata) but also by the metacognitive mechanisms that select, monitor, and evaluate the resources to be used. The next section discusses how metacognition influences knowledge organization for mathematical problem solving.

The Role of Metacognition

Metacognition refers to one's knowledge concerning one's own cognitive processes and products or anything related to them. It refers not only to one's awareness of cognitive processes but also to the self-monitoring, regulation, evaluation, and direction of the processes (Flavell, 1976). In the past few years, some attention has been paid to the nature and importance of metacognitive influences on mathematical problem solving (c.f., Schoenfeld, Note 6; Silver, Branca, and Adams, 1980; Silver, Note 7). The present discussion focuses on aspects of metacognition that might influence encoding and retrieval of information in the area of mathematical problem solving.

When faced with a problem, the solver may choose to employ various cognitive strategies to assist the solution. These decisions are essentially metacognitive in nature and influence profoundly the way in which the problem information is encoded and what other information is retrieved from memory. For example, a decision to use an elaboration strategy—to think of how the given problem is related to previously solved problems or previously learned material—will influence both the encoding and the retrieval of possibly useful information. Similarly, decisions to examine special cases, to draw a picture or a diagram, to restate the problem, or to solve a simpler problem can each influence the solver's processing of problem information. If we adopt a metacognitive perspective, we can view many of Polya's (1957) heuristic suggestions as metacognitive prompts that can profoundly influence a problem solver's information-processing behavior.

Decisions to employ certain cognitive strategies are influenced by one's beliefs and values. One is unlikely to examine cases and search for a pattern unless one *believes* that such a strategy has a reasonable chance of success. Recent work in cognitive anthropology and sociology has demonstrated that cultural belief systems can profoundly influence memory, perception, and cognition. Many components of a person's belief system or world view are metacognitive in nature, derived from sets of cognitive experience or passed on through the processes of acculturation and socialization. In the context of mathematical problem solving, a person's beliefs about schooling, learning, and problem solving in general and beliefs about mathematics and mathematics problem solving in particular can act as powerful guides in the encoding and retrieval of mathematical material.

Consider, for example, many of the characteristics of mathematically capable students: 1. generalized memory for mathematical relationships, type characteristics, schemes of arguments and proofs, methods of proofs, methods of problem solving, and principles of approach, 2. formalized perception of mathematical material, grasping the formal structure of a problem, and 3. rapid and broad generalization of mathematical objects, relations, and operations (Krutetskii, 1976). A person who *believes* that there is an underlying structure to mathematics and that this structure is more important than the surface details will approach the study of mathematical material quite differently than a student who does not hold this belief. If this belief is a component of one's mathematical "world view," then the characteristics described by Krutetskii follow quite naturally.

Other components of a mathematical belief system which may have important implications for how one approaches mathematical problems would include: 1. the belief that there is usually more than one way to solve a problem, 2. the belief that two different methods to solve a problem should each result in the same correct solution, and 3. the belief that there exists a most concise and/or clear way to present a problem or its solution. Naturally, many other beliefs could be added to the list.

The study of mathematical belief systems and their influence on problem-solving behavior is a fertile area for research. In particular, belief systems may prove useful in explaining strategy selection or non-selection, perseverance or non-perseverance, and feelings of satisfaction or dissatisfaction in problem-solving episodes.

A full consideration of belief systems and their influence on mathematical problem solving would also require examining the beliefs of the teacher. The work of George Shirk (1972) and Alba Thompson (Note 8) demonstrates that teachers' subjective theories about mathematics, mathematics learning, and mathematics teaching can influence profoundly the nature of classroom interaction. Mathematical material for classroom instruction is chosen by teachers according to their beliefs about mathematics learning and teaching. In this way, they may influence the availability and organization of mathematical knowledge for the student.

Several interesting research questions arise. What beliefs are held by experienced mathematics teachers concerning the nature and importance of mathematical problem solving? What beliefs are held by those reputed to be excellent teachers of problem solving? What beliefs are held by teachers participating in a study on the effectiveness of a particular method of problem-solving instruction? To what extent are the beliefs of teachers compatible with the implicit or explicit philosophy of the curriculum they teach?

Beliefs and values can also influence research on problem solving. Concern about the influence of a researcher's beliefs on the subjects in instructional studies has often led to the imposition of arbitrary "controls" to neutralize their effects. But instead of trying to neutralize their effects, we should begin to study systematically the ways in which belief systems influence problem-solving performance, the teaching and learning of problem solving, and research on these topics.

Summary

This paper has discussed several influences on knowledge organization and accessibility for mathematical problem solving. It has examined the role of certain cognitive factors (viz., problem schemata and elaboration), and certain metacognitive factors (viz., strategy selection and belief systems). The time has come to tackle the fundamental issues of knowledge organization and accessibility and to develop theories that use available formulations from psychology while reflecting the unique character of mathematics.

Reference Notes

1. Mayer, R. E. *Recall of Algebra Story Problems.* Unpublished manuscript, University of California, Santa Barbara, Department of Psychology, Series in Learning and Cognition, Report No. 80-5, 1980.
2. Heller, J. I., and Greeno, J. G. *Semantic Processing in Arithmetic Word Problem Solving.* Paper presented at the Midwestern Psychological Association, 1978.
3. Riley, M. S., and Greeno, J. G. *Importance of Semantic Structure in the Difficulty of Arithmetic Word Problems.* Paper presented at the Midwestern Psychological Association, 1978.
4. Briars, D., and Larkin, J. *An Integrated Model of Skill in Solving Elementary Word Problems.* Paper presented at AERA annual meeting, Los Angeles, CA, April 1981.
5. Mayer, R. E. *Elaboration Techniques and Advance Organizers That Affect Technical Learning.* Paper presented at AERA annual meeting, San Francisco, CA, 1979.
6. Schoenfeld, A. H. *Episodes and Executive Decisions in Mathematical Problem Solving.* Paper presented at AERA annual meeting, Los Angeles, CA, April, 1981.
7. Silver, E. A. *Thinking About Problem Solving: Toward an Understanding of Metacognitive Aspects of Mathematical Problem Solving.* Paper presented at The Conference on Thinking, Suva, Fiji, January 1982.
8. Thompson, A. G. *Teachers' Conceptions of Mathematics and Mathematics Teaching.* Doctoral dissertation, University of Georgia, in progress.

References

Anderson, R. C., Reynolds, R. E., Schallert, D. L., and Goetz, E. T. "Frameworks for Comprehending Discourse." *American Educational Research Journal,* 1977, *14,* 367-381.
Ausubel, D. P. *Educational Psychology: A Cognitive View.* New York: Holt, Rinehart and Winston, 1968.
Bransford, J. D., Barclay, J. R., and Franks, J. J. "Sentence Memory: A Constructive Versus Interpretive Approach." *Cognitive Psychology,* 1972, *3,* 193-209.
Davis, R. B., Jockusch, E., and McKnight, C. "Cognitive Processes in Learning Algebra." *The Journal of Children's Mathematical Behavior,* 1978, *2,* 10-320.
Davis, R. B., and McKnight, C. "Modeling the Processes of Mathematical Thinking." *The Journal of Children's Mathematical Behavior,* 1979, *2,* 91-113.
Flavell, J. H. "Metacognitive Aspects of Problem Solving." *In* L. B. Resnick (Ed.), *The Nature of Intelligence.* Hillsdale, NJ: Lawrence Erlbaum Associates Inc., 1976.

Hinsley, D. A., Hayes, J. R., and Simon, H. A. "From Words to Equations —Meaning and Representations in Algebra Word Problems." *In* M. Just and P. Carpenter (Eds.), *Cognitive Processes in Comprehension.* Hillsdale, NJ: Lawrence Erlbaum Associates Inc., 1977.

Krutetskii, V. A. *The Psychology of Mathematical Abilities in Schoolchildren,* J. Kilpatrick and I. Wirszup (Eds.). Chicago: University of Chicago Press, 1976.

Loftus, E. F., and Suppes, P. "Structural Variables that Determine Problemsolving Difficulty in Computer-assisted Instruction. *Journal of Educational Psychology,* 1972, *63,* 531-542.

Luchins, A. S. "Mechanization in Problem Solving. *Psychological Monographs, 1942, 54(6),* Whole No. 248.

Mandler, J. M., and Johnson, N. S. "Remembrance of Things Parsed: Story Structure and Recall." *Cognitive Psychology,* 1977, *9,* 111-151.

Mayer, R. E. "Integration of Information During Problem Solving Due to a Meaningful Context of Learning." *Memory and Cognition,* 1976, *4,* 603-608.

Mayer, R. E. "The Sequencing of Instruction and the Concept of Assimilation-to-schema." *Instructional Science,* 1977, *6,* 369-388.

Mayer, R. E., and Greeno, J. G. "Structural Differences Between Learning Outcomes Produced by Different Instructional Methods. *Journal of Educational Psychology,* 1972, *63,* 165-173.

Polya, G. *How to Solve It* (2nd ed.). New York: Doubleday & Co. Inc., 1957.

Polya, G. *Induction and Analogy in Mathematics.* Princeton, NJ: Princeton University Press, 1973.

Reder, L. M. "The Role of Elaboration in the Comprehension and Retention of Prose." *Review of Educational Research,* 1980, *50,* 5-53.

Rohwer, W. D. "An Introduction to Research on Individual and Developmental Differences in Learning." *In* W. K. Estes (Ed.), *Handbook of Learning and Cognitive Processes, Vol. 3: Approaches to Human Learning and Motivation.* Hillsdale, NJ: Lawrence Erlbaum Associates Inc., 1976.

Rohwer, W. D. "Images and Pictures in Children's Learning." *Psychological Bulletin,* 1970, *73,* 393-403.

Shirk, G. B. *An Examination of Conceptual Frameworks of Beginning Mathematics Teachers.* Unpublished doctoral dissertation, University of Illinois, Urbana-Champaign, 1972.

Silver, E. A. "Student Perceptions of Relatedness Among Mathematical Verbal Problems." *Journal for Research in Mathematics Education,* 1979, *10,* 195-210.

Silver, E. A., Branca, N. A., and Adams, V. M. "Metacognition: The Missing Link in Problem Solving?" *In* R. Karplus (Ed.), *Proceedings of the Fourth International Conference for the Psychology of Mathematics Education.* Berkeley, CA, 1980.

Silver, E. A., and Smith, J. P. "On Making Effective Use of the Advice, Think of a Related Problem." *In* S. Krulik (Ed.), *Problem Solving in School*

Mathematics. Reston, VA: National Council of Teachers of Mathematics, 1980.

Simon, H. A. "Problem Solving and Education." *In* D. T. Tuma and F. Reif (Eds.), *Problem Solving and Education*. Hillsdale, NJ: Lawrence Erlbaum Associates Inc., 1980.

Spiro, R. J. "Remembering Information from Text: The 'State of Schema' Approach." *In* R. C. Anderson, R. J. Spiro, and W. E. Montague (Eds.), *Schooling and the Acquisition of Knowledge*. Hillsdale, NJ: Lawrence Erlbaum Associates Inc., 1977.

Thorndyke, P. W., and Hayes-Roth, B. "The Use of Schemata in the Acquisition and Transfer of Knowledge." *Cognitive Psychology*, 1979, *11*, 82-106.

Thorndyke, P. W., and Yekovich, F. R. "A Critique of Schema-based Theories of Human Story Memory." *Poetics*, 1980, *9*, 23-49.

Weinstein, C. E., Underwood, V. L., Wicker, F. W., and Cubberly, W. E. "Cognitive Learning Strategies: Verbal and Imaginal Elaboration." *In* H. F. O'Neil and C. D. Spielberger (Eds.), *Cognitive and Affective Learning Strategies*. New York: Academic Press Inc., 1979.

Yaroshchuk, V. L. "A Psychological Analysis of the Processes Involved in Solving Model Arithmetic Problems." *In* J. Kilpatrick and I. Wirszup (Eds.), *Soviet Studies in the Psychology of Learning and Teaching Mathematics* (Vol. 3). Stanford: School Mathematics Study Group, 1969. (Originally published, 1957.)

Some Thoughts on Problem-solving Research and Mathematics Education

Alan H. Schoenfeld

As I tried a number of earlier versions of this paper, I came to realize that two questions lay behind all of the issues with which I was grappling. First, why do we teach mathematics? And second, why do we do research in problem solving?

Since these questions sound pretentious if not downright silly, I have some explaining to do. Part I of this paper presents the four themes that vied for center stage in earlier versions of it.

I. I believe that most instruction in mathematics is, in a very real sense, deceptive and possibly fraudulent. These are harsh words. Here are three examples to justify them.

Example 1

"Word problems" are one of the major focal points of mathematics instruction in the elementary schools. Typical of such problems at the lower grade levels is "John had eight apples. He gave three to Mary. How many does John have left?"

Much instruction on how to solve such problems is based on the "key word" algorithm, where the student makes his choice of the appropriate arithmetic operation by looking for syntactic cues in the problem statement. For example, the word "left" in the problem given above serves to tell the student that subtraction is the appropriate operation to perform. At the research presessions to the 1980 annual NCTM meeting, these two facts were reported:

1. In a widely used elementary textbook series, 97 percent of the problems "solved" by the key-word method would yield (serendipitously?) the correct answer.
2. Students are drilled in the key-word algorithm so well that they will use subtraction, for example, in almost any problem containing the word "left." In the study from which this conclusion was drawn, problems were constructed in which the appropriate operations were addition, multiplication, and division. Each used the word "left" conspicuously in its statement and a large percentage of the students subtracted. In fact, the situation was so extreme that many students chose to subtract in a problem that began "Mr. Left . . . ."

Example 2

I don't know about nationwide enrollment figures, but I suspect that those for Hamilton College are typical, if not low: some 60 percent of Hamilton's students study calculus, but fewer than 10 percent of them go on to take more advanced mathematics. At the University of Rochester 85 percent of the freshman class takes calculus, and many go on. Roughly half of our students see calculus as their last mathematics course. Most of these students will never apply calculus in any meaningful way (if at all) in their studies, or in their lives. They complete their studies with the impression that they know some very sophisticated and high-powered mathematics. They can find the maxima of complicated functions, determine exponential decay, compute the volumes of surfaces of revolution, and so on. But the fact is that these students know barely anything at all. The only reason they can perform with any degree of competency on their final exams is that the problems on the exams are nearly carbon copies of problems they have seen before; the students are not being asked to think, but merely to apply well-rehearsed schemata for specific kinds of tasks. Tim Keiter and I studied students' abilities to deal with pre-calculus versions of elementary word problems such as the following:

> An 8-foot fence is located 3 feet from a building. Express the length L of the ladder which may be leaned against the building and just touch the top of the fence as a function of the distance x between the foot of the ladder and the base of the building.

We were not surprised to discover that only 19 of 120 attempts at such problems (four each for 30 students) yielded correct answers, or that only 65 attempts produced answers of any kind. We were surprised, however, to discover that much of the students' difficulty came not from the "problem solving" part of the process (setting up and solving systems of equations) but from the *reading* part of it.

> Fifty-eight protocols were obtained from randomly selected calculus students who were asked to rewrite problem statements "more understandably." Of these, 5 simply rewrote the problem verbatim. The 53 remaining rewrites tell a sorry tale: 5 (9.4%) included information which directly contradicted the input, and 11 (20.4%) contained information that was so confused as to be unintelligible; 2 students (4%) made both kinds of errors. This information is the more striking since two-thirds of these students were to write simple declarative sentences, if possible, to make their task simpler. Thus before they would normally have put pen to paper, a quarter of the 53 students had already seriously garbled or completely misinterpreted the problem statement. None of those students ever got an answer to the problem (Keiter, Note 1).

Those students had already "covered" word problems in their calculus classes.

Example 3

I taught a problem-solving course for junior and senior mathematics majors at Berkeley in 1976. These students had already seen some remarkably sophisticated mathematics. Linear algebra and differential equations were old hat. Topology, Fourier transforms, and measure theory were familiar to some. I gave them a straightforward theorem from plane geometry (required when I was in tenth grade). Only two of eight students made any progress on it, one of them by using arc length integrals to measure the circumference of a circle (Schoenfeld, 1979). Out of the context of normal course work, these students could not do elementary mathematics.

In sum: all too often we focus on a narrow collection of well-defined tasks and train students to execute those tasks in a routine, if not algorithmic fashion. Then we test the students on tasks that are very close to the ones they have been taught. If they succeed on those problems, we and they congratulate each other on the fact that they have learned some powerful mathematical techniques. In fact, they may be able to use such techniques mechanically while lacking some rudimentary thinking skills. To allow them, and ourselves, to believe that they "understand" the mathematics is deceptive and fraudulent.

II. The mathematics education community has isolated itself from psychological and other research in problem solving. Mathematics education is a young and unsettled discipline. The case can be made that the phoenix of a "process-oriented approach" to math-education problem-solving research rose from the ashes of the statistical approach in the mid- and late 1960s; we are now in our adolescence, and experiencing growing pains. Yet the community has made life much harder for itself than it has had to. In a recent book on problem solving, five of the nine dissertation studies presented dealt with students in the fourth- through seventh grades (Harvey and Romberg, 1980). However, the extensive literature of developmental psychology was all but ignored—a 31-page set of references did not include a single work by Piaget. Similarly, a variety of studies in mathematics education have used protocol analysis and agonized over the effects of verbalization on problem-solving performance. This topic has been studied extensively in the psychological literature (Ericsson and Simon, 1980). There is no need for us to reinvent that particular methodological wheel, or any of a number of others. To put it bluntly, it may be impossible to do "state of the art" work in math-education problem-solving research without a solid background in the relevant psychological research. For example, the detailed process models offered by information-processing psychologists and the results of studies on verbalization are essential for my own work. They provide a foundation for it that is not available within mathematics education.

These comments are not meant in any way to suggest that mathematics education should become an adjunct to cognitive psychology, or even consider adopting its ideas and perspectives wholesale. It seems to me that there are significant and dangerous implications present in some of the theoretical underpinnings of modern cognitive psychology, especially in those of information-processing psychology. I will list a few points of concern here, and discuss them at greater length below.

Among points of concern are the following. There is the phenomenon of methodologically-induced focus: one tends to examine those aspects of things that our methodologies will illuminate, and to de-emphasize or ignore those that are not compatible with them. "Models" of the problem-solving process can cause difficulty in at least two ways. They may ignore aspects of the problem-solving process that cannot (currently) be modeled or are incompatible with the current modeling perspective. It will be interesting to watch how information processing comes to grips with issues of metacognition, for example. Also, there is the danger that the models can be taken too seriously, as explanations of cognitive performance. In the sense just described, they may be reductive. In the sense that they are only potential explanations of performance, a particular model may be dead wrong! (See the example of the student teachers who "got the bug" in Brown and Burton, 1978.) We must remember that models of experts and novices are just that; the extrapolation from the models back to real people must be done with care. There is the danger that, better armed with procedures for decomposing certain kinds of cognitive tasks, we will misuse them and become more sophisticated at perpetuating the kinds of deceptions I mentioned in Theme I. Also, there is currently a fair amount of confusion about what it means to be an "expert" or "novice." Let me paraphrase a comment made to me by John Seely Brown: there can be a significant difference between "expertise" and the ability to perform well in a domain.

III. The world of problem solving is small and possibly incestuous. A few years ago I asked a number of colleagues involved in problem-solving research if they had collections of good problems. Among the people who responded was Ed Silver. I was familiar with virtually all of the problems Ed sent. Most were from Polya and other standard sources. A few were problems I had created, which (I believe) had made their way to Ed via John Lucas.

The point is that the mathematics education community has a very narrow perspective on what "problem solving" means. One need only look at the *1980 NCTM Yearbook* to see that virtually all the authors discuss the same kinds of "nonroutine" problems, if not the same problems themselves! (I was asked to change some of the examples in my article because they duplicated the examples in other articles.) Ed commented then that he was concerned about the incestuous nature of the community: a small number of researchers shared interests and problems, and all seemed to be investigating this narrow collection, which went by the name of "problem solving." I fear that his comment may be accurate. For example, the students who have learned, in algorithmic fashion, to "substitute n-1, 2, 3, 4 for an integer parameter and look for a pattern" may be solving difficult problems, but are they doing problem solving?

IV. There is a difference between my choices of problems and my notion of expert, and the standard choices of problems and notion of "expert." In a recent conversation, Dick Lesh pointed out that the tasks used in my recent problem-solving studies are not the standard "non-standard" problems, and that my "experts" display markedly different (and often remarkably less proficient) behavior than most "experts." For example, in my research I give both students and colleagues problems that are either unfamiliar to them or are from domains they studied long ago. A particular favorite of mine is the following problem:

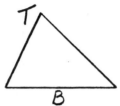

You are given a fixed triangle T with base B. Show that it is always possible to construct, with straightedge and compass, a straight line parallel to B such that the line divides T into two parts of equal area. Can you similarly divide T into five parts of equal area?

This problem is difficult for most students and, in fact, has proven difficult for some of my colleagues. The solution provided by one problem solver, GP, was derided by another, JTA, as being stupid and clumsy. Yet I chose GP's "stupid and clumsy" solution for analysis (Schoenfeld, in press) as an expert protocol, and found JTA's clean solution of little interest. So my tests of problem solving do not examine what I have just taught students, and my "experts" appear inexpert by standard criteria. This is not a matter of perversity, but one of perspective. It is tied to the first three themes and to the two questions with which I opened this paper. I would like to give my personal answers to these questions, and then discuss Themes 2 and 4 from that perspective.

Questions and Some Subjective Answers

Why do we teach mathematics? Not because mathematics is useful, although it is: our curricula reveal that. But how often does one need to determine how rapidly a person could row in the absence of a current, if it takes so long to row with a constant current and so long to row against it? Or for that matter, how often does one need to use a trigonometric identity, or virtually anything from Euclidean geometry, or to calculate the volume of a surface of revolution? Mathematics can be applied to the real world, although we do a rather poor job of teaching our students to do it. We do an even poorer job of selecting potentially useful and meaningful problems for our students to master. But that is only part of the story.

Other parts have to do with the scope and power of the discipline. It is a massive intellectual achievement, and should be appreciated even if not used. It is as well a marvelously aesthetic discipline, and it would be nice to have our students appreciate it for that. But in my opinion the single most important reason to teach mathematics is that it is an ideal discipline for training students how to *think*. I will try to characterize "thinking" in more detail later, but for now the usual sense of the word will suffice. Mathematics is a discipline of clear and logical analysis that offers us tools to describe, abstract, and deal with the world (and later, worlds of ideas) in a coherent and intelligent fashion. Our goal as teachers should be to teach students to use mathematics that way.

For example, the calculus version of the pre-calculus "ladder and fence" problem given earlier is ludicrous. If one ever did need to solve such a problem, it could probably be best accomplished by rough empirical methods. But it is worthwhile having students work on such problems. To solve this problem the student must extract the relevant information from the text, create an accurate diagram with the appropriate symbolic notation, establish goals and subgoals, and seek (from memory) the relevant information that will allow the goals and subgoals to be achieved. Further, all of this must be done with reasonable efficiency, and students must learn that as well. To the degree that this problem serves as a vehicle for developing those skills, it is worthwhile. Taken in and of itself, or as an exemplar of a class of problems, it is of questionable value. The same is true of much of the mathematics we teach.

Why do research in problem solving? From my perspective, it is so that we can better understand what constitutes productive thinking skills, so that in turn we can be more successful in teaching students to think. It is not easy to define *Thinking*. (I shall use the uppercase T to distinguish Thinking from the ordinary associations of the word.) Here are some examples of what it is not. A mathematician is not Thinking when he uses the quadratic formula. That should come as no surprise, since the application of the formula is algorithmic. But most probably he has no need to Think when he solves the pre-calculus problem given earlier. That problem is completely routine for college mathematics teachers, as are virtually all calculus problems. Even if he has not worked a problem isomorph of it before, the mathematician would in all likelihood be able to crank out a solution to it with as much ease as he could factor the expression $(6x^2+17x+12)$. If you were to observe or attempt to model his performance on that type of problem, you would be a spectator to a demonstration of domain-specific proficiency, but you would not see whatever it is that accounts for his problem-solving skill. The same is true for virtually all schema-driven solutions, including "heuristic" solutions to "non-routine" problems (if the "expert" has access to the schema).

To examine what accounts for expertise in problem solving, you would have to give the expert a problem for which he does *not* have access to a solution schema. His behavior in such circumstances is radically different from what you would see when he works on routine or familiar "non-routine" problems. On the surface his performance is no longer proficient; it may even seem clumsy. Without access to a solution schema, he has no clear indication of how to start. He may not fully understand the problem, and may simply "explore" it for a while until he feels comfortable with it. He will probably try to "match" it to familiar problems, in the hope it can be transformed into a (nearly) schema-driven solution. He will bring up a variety of plausible things: related facts, related problems, tentative approaches, etc. All of these will have to be juggled and balanced. He may make an attempt at solving it in a particular way, and then back off. He may try two or three things for a couple of minutes and then decide which to pursue. In the midst of pursuing one direction he may back off and say "that's harder than it should be" and try something else. Or, after the comment, he may continue in the same direction. With luck, after some aborted attempts, he will solve the problem.

Does that make him, at least in that domain, a bad problem solver? I think not. In all likelihood someone proficient in that domain (i.e., someone who knows the right schemata) could produce a solution that puts his to shame. But that isn't the point at all. The question is: How effectively did the problem solver utilize the resources at his disposal?

One of the most impressive protocols I have ever seen is the "stupid and clumsy" solution produced by expert GP to the problem given in Theme 4 (see Schoenfeld, in press). The protocol is five single-spaced pages long (20 minutes), and a detailed analysis takes longer. GP has no idea what "makes the problem tick," and remembers less of his plane geometry than my college freshmen, who have studied the subject much more recently. He generates enough potential sources of "wild goose chases" in his protocol to mislead an army of problem solvers. But unlike my students, he manages not to be misled. His protocol is a *tour de force* of metacognition; rarely do more than 15 seconds elapse between comments on the state of his own knowledge and the state of the solution. While he is fertile in generating potential solution paths, he is also ruthless in pruning them. With less domain-specific knowledge at his disposal than most of my students had, he managed to solve a problem that left all of them stymied. Therein lies his "expertise." It is not simply the possession of schemata that allows him to solve problems with dispatch, although that is an important component of his competence. It is instead the ability to deploy the resources at his disposal so he can make progress while others wander aimlessly.

Implications

One point I wish to stress is that proficiency (the possession of a large number of schemata for dealing with generic classes of tasks in a domain) should not be confused with expertise. There are dangers in confusing the two.

In the short run, proficiency models (which is what virtually all "expert" models have been) are useful. It is worthwhile, for example, to develop schemata for elementary word problems that are mathematically and psychologically valid, and accessible to school children. A system of instruction based on these would obviously be preferable to the "key word" system, which uses illegitimate means to achieve what may be "rigged" performance objectives. Properly interpreted and used, information-processing models of competent performance are valuable. In any field, cleaner instruction resulting in improved performance can hardly be unwelcome.

There are difficulties at both microscopic and macroscopic levels in how to interpret and use performance models. First, at the microscopic level there is the problem that modeling can, at times, be an end in itself. It should instead serve as the beginning for a new set of inquiries. There are now, in a number of domains, production system models that not only simulate and predict performance but can be modified to "improve" or "grow." In some very clever work now being done at Carnegie-Mellon University (Briars and Larkin, note 2), a series of nested production models has been developed for solving elementary word problems. A running program performs at a level consistent with the performance of kindergarteners. Adding one production to the system (and some minor modifications) results in performance like that of first graders, and adding one more results in performance like that of second graders. Performance predicted by the models agrees very well with empirical data, and the models serve to unify collections of empirical data and to provide a framework from which to make predictions. But do the processes in the models really reflect the cognitive processing in the children who are being modeled? The theory suffices to make predictions, without the implementation of the program. One has the feeling (in this case confirmed by conversation with one of the authors) that the programs are important because the authors believe the processes in the program *are,* at some level, the processes in the minds of the students. This belief is fine, if it is considered a hypothesis to be tested. If left unquestioned, it is severly reductive and can have dangerous consequences. Moreover, it should be recognized that the hypothesis, even if correct, now gives rise to the *real* question: Just what happens during a full year of a child's development that results in the addition of *one* production to his word-problem "program"?

I will find it very interesting to keep an eye on this particular line of research. In a recent conversation, Diane Briars told me that when solving problems, the students would often encounter contradictions between their intuitions and the processes they had been taught to use. They would say things like, "I know it ought to be larger, but I'm supposed to use subtraction." Most often they would succumb to their training. So far as I'm concerned, the metacognitive aspects of this process—the generation of the students' intuitions and the means that the students use to resolve the conflicts—lie at the heart of their performance. The "purely cognitive" aspects of their performance, which have been modeled, tell a critically important part of the story. However, the models do not take the metacognitions into acount—they cannot, at present. To elaborate the models means to ignore an important part of psychological reality, but to deal with that reality means to abandon current methodology. Where does one go next?

These comments are not to be taken as an indictment of this study. I chose to discuss it because it is a good study, relevant to some of the themes I raised earlier. But the questions I have just raised apply to most artificial intelligence studies, and are rarely raised (at least in print or in my company) by those who create them. At the recent AERA (American Educational Research Association) meeting, Lauren Resnick characterized much artificial intelligence work as "art in the service of science." We must make certain that it does indeed serve.

The second and much more perilous difficulty lies in interpreting and using performance models at the macroscopic level. There is a very serious danger when proficiency and expertise are confused, and expertise is defined as proficiency. Thinking (with the uppercase "T") is then defined out of existence, or banished to irrelevancy. The situation is exacerbated by a kind of "proof is in the pudding" argument that goes something like this: "We have produced programs that operate successfully without any need for construct X. Further, people have tried to construct programs based on construct X and failed. Therefore, construct X, even if it does exist, is at best of minor importance." This particular statement was made to me about heuristics, but could also have been made about metacognition, Thinking, or any of a number of potentially important domains of inquiry. Most theoretical artificial intelligence and information-processing work these days is done, *de facto*, along proficiency model lines: "experts" always seem to be performing routine tasks, and theoretical work now focuses on models of productive thinking via scripts or schemata. If the traditional evolutionary pattern holds, applied research will follow suit, and so will educational research and development. During the height of behaviorism, certain "mental constructs" were *déclassé*, and to be shunned at all costs. Let us not make similar mistakes about Thinking in a world dominated by proficiency models. That approach can only deflect us from the global goals we have in teaching and research.

Final Comments

The mathematics education community cannot afford to ignore the psychological research on problem solving. But it cannot afford to swallow it whole, either. Mathematics educators, I think, have had their hearts in the right place but have lacked the methodological tools that allowed for substantive and rigorous inquiries into problem solving. Many such tools have been developed by the psychological community, and much of our work will be second rate at best if we do not take advantage of them. As I mentioned above, it would be impossible for me to do my own work without the support of research into the effects of verbalization on problem-solving performance or the substantive ideas underlying the modeling of cognitive processes.

There is a great deal more to problem solving than is currently being modeled. I personally am convinced that metacognitions play a tremendous role as "driving forces" in cognitive performance, and that much more research needs to be done in exploring them. They have cropped up in various ways in this paper. They include the monitoring and assessment strategies that students lack and experts possess, allowing the former to go off on "wild goose chases," while the latter works efficiently (Schoenfeld, in press); they include the intuitions against which the progress or plausibility of a solution is gauged, and the means by which such conflicts are resolved; and they include both the conscious and unconscious belief systems that may determine the approaches people take to certain problems. These areas barely have been touched upon, and need much more research. They are just some of many that we will discover in an open-ended, open-minded quest for knowledge and understanding. We are beginning to make progress, and can hope to see more.

Reference Notes

1. Keiter, T. *Solving Calculus Word Problems: Behavior and Learning.* Senior Fellowship thesis, Hamilton College, 1981.
2. Briars, Diane J. and Larkin, Jill H. *An Integrated Model of Skill in Solving Elementary Word Problems.* Paper presented at the 1981 annual AERA meeting, Los Angeles, April, 1981.

References

Brown, J. S. and Burton, R. R. "Diagnostic Models for Procedural Bugs in Basic Mathematical Skills." *Cognitive Science* 2, 155-192 (1978).

Ericsson, K. A. and Simon, H. A. "Verbal Reports as Data." *Psychological Review,* 1980, 87(3), 215-251.

Harvey, J. G. and Romberg, T. A. *Problem Solving Studies in Mathematics.* Madison, Wisconsin: Wisconsin Research and Development Center for individualized schooling monograph series, 1980.

National Council of Teachers of Mathematics. *1980 Yearbook: Problem Solving in School Mathematics.* S. Krulik (Ed.), Reston, Va.: National Council of Teachers or Mathematics, 1980.

Schoenfeld, A. H. "Teaching Problem Solving in College Mathematics: the Elements of a Theory and a Report on the Teaching of General Problem Solving Skills." *In* R. Lesh, D. Mierkiewicz and M. Kantowski (Eds.), *Applied Mathematical Problem Solving.* Columbus, Ohio: ERIC/ SMEAC, 1979.

Schoenfeld, A. H. "Episodes and Executive Decisions in Mathematical Problem Solving." *In* R. Lesh and M. Landau (Eds.), *Acquisition of Mathematics Concepts and Processes.* New York: Academic Press, (in press).

Implications from Information-processing Psychology for Research on Mathematics Learning and Problem Solving*

Diane J. Briars

Information-processing psychology is currently one of the leading approaches to studying human cognition. Yet, it has had relatively little influence on most research on mathematics learning and problem solving. This is a result of a number of factors, not the least of which is that early information-processing research focused on verbal learning and solving puzzle-type problems, domains that appear to have little in common with mathematics. More recently, however, researchers with an information-processing perspective have started to investigate learning and problem solving in "real" domains like physics (Larkin, McDermott, Simon, and Simon, 1980), computer programming (Jeffries, Turner, Polson, and Atwood, 1981), and mathematics (Riley, Greeno, and Heller, in press; Briars and Larkin, Note 1; Greeno, 1978; Lewis, 1981; Neves and Anderson, 1981). The dominance of the information-processing approach in cognitive psychology, coupled with increasing numbers of information-processing studies of mathematical problem solving, makes it timely to ask: What new ideas and perspectives does information-processing psychology bring to the study of mathematics learning and problem solving?

Although specific results from information-processing studies of mathematical learning and problem solving are certainly important, the most potentially impactful ideas are the fundamental concepts underlying the approach. One distinguishing feature is that humans are characterized as information-processing systems. Another is the methodology developed to test information-processing theories, specifically the use of computer simulation models of human cognition. The first sections of this paper discuss these two fundamental aspects and their implications; the final section briefly presents some recent advances in the study of human problem solving from information-processing psychology.

People as Information-processing Systems

In information-processing psychology, people are viewed as entities that take in, process, and sometimes report information. Information is broadly

*This work was supported by NSF grant number 1-55035. Any opinions expressed are those of the author and do not reflect the views of the National Science Foundation.

defined, though usually it refers to some kind of symbol structure (Simon, 1978). The fundamental difference between this approach to the study of cognition and those preceding it (e.g., behaviorism) is the focus on the transformation of information and the processes by which that occurs, rather than on inputs and outputs (stimulus and responses) (Estes, 1978).

This focus makes questions about cognitive mechanisms that process and store information and the constraints they impose on learning and problem solving central issues. Consequently, initial information-processing research largely concerned the structure of human memory.* Memory is usually characterized as consisting of two different kinds of information stores, or alternatively, of two different ways of accessing information (Anderson, 1980).** *Short-term memory* (STM) retains information only temporarily; it is STM that "holds" an unfamiliar phone number for the few seconds between the time it is looked up and the time it is dialed. *Long-term memory* (LTM) retains information over a long period of time, serving as an individual's relatively permanent store of knowledge. The basic constraints on the information-processing system are, then, the capacity and storage and access speeds of its memories (Simon, 1979a).

All information is taken in by STM. Furthermore, this is the memory that "processes" or transforms all information—both new information from the environment and information recalled from LTM. As a result, STM is the major bottleneck of the information-processing system.

STM has relatively fast storage and retrieval: on the order of several hundred milliseconds (Simon, 1981). Its major constraint is its limited capacity. STM capacity is measured in "chunks," where a chunk is information that can be represented by a single symbol (Miller, 1956). For example, "dog" is a single chunk even though it consists of three letters because it is a symbol for a single entity, while "zcq" is three chunks for most people. The size of a chunk seems to depend on an individual's previous knowledge. For example, the eight digits 19411776 may consist of only two chunks, "1941" and "1776," because these numbers are well-known years in American history. On the other hand, the eight digits 70915296 would probably not be chunked as 7091 and 5296 unless these numbers had special significance for an individual.*** This sequence could be interpreted as eight chunks if each digit was considered separately, or as four chunks if the sequence was interpreted as seventy, ninety-one, fifty-two, ninety-six. Due to this dependence on previous knowledge, STM is usually described as holding

---

*Sensory and perceptual processes are not included in this discussion, though they are important elements of cognition. See Chase, 1978, and Turvey, 1978, for discussions of these other processes.

**This is a simplified description of memory, including only the features common to most models. See Anderson, 1976, Simon, 1976, and Crowder, 1976, for detailed discussions of different memory models.

***See Chase and Ericsson, 1981, for a discussion of chunking in expert memory (i.e., the ability to recall 80 random digits).

pointers to knowledge stored in LTM, rather than actually containing the chunks of information. Thus, STM capacity is defined in terms of the number of pointers it can retain. Estimates of adult STM capacity range from four (Simon, 1979a) to seven (Miller, 1956) chunks of information. Furthermore, there is some evidence that STM capacity increases with age (Case, 1978), though there is obviously an interaction between age and knowledge available to facilitate chunking.

In contrast, LTM does not appear to have limited capacity. Instead, its primary constraints are the time required to incorporate information into LTM and ability and time required to retrieve stored information. Time to initially enter information into LTM (transfer symbols from STM to LTM) has been estimated at 5 to 10 seconds per chunk and time to initially access a chunk at two seconds, with several hundred milliseconds needed to access chunks beyond the first. However, the more important constraint on retrieval is *ability* to access. Ability to retrieve information depends on three components: the contents of LTM itself, how this content is organized, and the route by which desired information can be accessed. A useful analogy is to consider the first two the "text" of a book and the last the book's index (Simon, 1979a).

Most information-processing models of memory have the same core description of LTM as node-link structures or networks. Each node stores a symbol and each node is connected to other nodes by links, which represent relations between nodes. For example, "apple," "cherry," and "grape" may each correspond to a memory node, and all may be linked to the "fruit" node by a superordinate relation, and may also be linked to each other. Further, a single node may refer to a complex symbol (e.g., a configuration of chess pieces); such a node could be "unpacted" or decomposed into its elements by considering the nodes linked to it by an "element of" relation. The "index" to LTM is most often characterized as a discrimination net (e.g., EPAM [Feigenbaum, 1961; Simon and Feigenbaum, 1964]; SAL [Hintzman, 1968], and MAPP [Simon and Gilmartin, 1973]). This net is a decision tree that successively checks features of stimuli or desired stimuli to determine an appropriate node.

In addition to this basic structure, constructs describing more complex organization of knowledge have been proposed. One of the most widely applied constructs is the concept of a memory schema and related ideas of scripts and frames (Rumelhart, 1975; Schank and Abelson, 1977). A schema is a collection of memory structures that describes the typical features and properties of the concept it represents. More specifically, it "is used to organize the components of specific experience and to expand the representation of an experience or message to include components that are not specifically contained in the experience, but that are needed to make the representation coherent and complete in some important sense" (Greeno, 1980, p. 718-719). For example, consider the activity "going to the movies." For most people, this activity is characterized by some typical features: buying tickets, a concession stand that sells popcorn and candy, a ticket-taker, watching previews of coming attractions, watching a film, etc. This knowledge can be represented by a "going to the movies" schema. Thus, a schema provides a prototypical description of the concept it represents that can be used to interpret a range of specific instances of that concept, and also to infer features of the concept that are not explicitly described.

Another key idea is the distinction between declarative and procedural knowledge in LTM (Anderson, 1980). As their names imply, declarative knowledge consists of facts—knowing that; procedural knowledge consists of skills—knowing how. This distinction between types of knowledge is not completely clean. Nonetheless, declarative knowledge can usually be expressed verbally, whereas procedural knowledge cannot. Most physical skills (riding a bicycle, hitting a tennis ball) are good examples of procedural knowledge that is difficult to describe verbally. The importance of this distinction is that each type of knowledge is hypothesized to be represented differently in LTM, and consequently, to have its own processing advantages and disadvantages. Declarative knowledge is thought to be more flexible in its application, more easily analyzed, and more easily modified, although slow and clumsy to apply. Procedural knowledge can be applied more quickly and easily, especially as the procedure becomes more automatic. It can handle variables more easily, but cannot be inspected (reflected upon) or modified easily (though it can be replaced). Procedural knowledge is more narrow therefore in its applicability. (For a detailed discussion of declarative and procedural knowledge and their implications, see Neves and Anderson, 1981.)

Implications for Research on Mathematics Learning and Problem Solving

Few of the characteristics of memory described above are even mentioned in most research on mathematics learning and problem solving. Although the quantitative estimates of encoding and retrieval times are of general theoretical interest, the qualitative aspects of this description—limited STM capacity, chunking information, and organization of knowledge in memory—appear most relevant for research on mathematics learning and problem solving.

STM Limitations

External aids can compensate for limited STM capacity in some situations (e.g., arithmetic calculations such as 3817 x 4723), but not in all. STM limitations seem to influence most choice of and ability to use particular problem-solving strategies. This is especially true for young children, who are thought to have a more limited STM capacity. Although STM or processing capacity has not been systematically considered in evaluating the development of children's mathematical concepts, there are several examples in the literature which point to its potential importance.

Case (1978) designed an instructional sequence on solving missing addend problems (e.g., 4 + __ = 7) that took into consideration children's limited processing capacity. He found that kindergarteners receiving this instruction much more successfully solved missing addend problems than those who received standard instruction (80 percent versus 10 percent of the children reached criterion in the same length of time). Memory capacity may also be an important constraint on children's abilities to use certain counting strategies in solving elementary addition and subtraction problems. Fuson, Richards, and Briars (in press) found a developmental sequence in the acquisition of counting skills in which the more difficult tasks required retention of an additional piece of information or extra processing (e.g., counting up from a given number was easier than counting up from that number and stopping at a specified number). One explanation for this sequence is that these harder tasks require processing capacity that exceeds that of the younger children. Finally, Romberg and Collis (1981) examined four- through eight-year-olds' performance on elementary addition and subtraction word problems as a function of their processing capacity. They found a significant increase in problem-solving skill as processing capacity increased. In addition, children with greater processing capacity relied less on direct modeling strategies and more on counting strategies and other mental operations than children with less processing capacity.

These studies are a first step in examining the role of processing capacity in mathematics learning and more research is certainly needed. These results are clouded by the existing problems with reliable measurement of working memory capacity, but better measures are being developed (Case, 1978). Nonetheless, processing capacity appears to be an important variable, especially in studies of young children's mathematical concepts and problem-solving skills.

Chunking Information and Knowledge Organization

As described earlier, there is a close relationship between how information is chunked in STM and how it is organized in LTM. A chunk is the basic element into which information is decomposed and represents a single memory node, though this node may be considered a set of subnodes. How incoming information is chunked may be an important variable in research on mathematical problem solving, possibly providing a more precise way of measuring differences between high- and low-ability students' organization of mathematical knowledge.

A number of recent studies have investigated differences in students' organization of mathematical knowledge and are discussed in Silver's chapter in this volume. The prime result of these studies is that higher ability students tend to organize their knowledge into conceptually rich mathematical schema, while less able students do not. This may also mean that more able students are encoding the problem in larger chunks (i.e., relational chunks) than the less able students. Because STM chunks are pointers to nodes in LTM, this description is consistent with the current hypothesis that more able students have knowledge organized differently, that chunking is possible because of the way knowledge is stored in memory. The relation between chunks and knowledge organization suggests that a recall paradigm similar to that used in the chess studies of de Groot (1965) and Chase and Simon (1979) could be used to assess individual differences in knowledge organization. Although some work has been done on long-term recall (Silver, 1981), little has been done in the way of immediate recall of problem information, especially when the stimuli were complex. Such a paradigm may be a useful alternative to the problem-similarity paradigm that has been extensively used in past research on knowledge organization.

Information-processing Methodology

The other distinguishing feature of information-processing psychology is the construction of computer simulation models of behavior. The rationale underlying use of these models is that people are processors of information that is in symbolic form—they are complex symbol processors. As such, they possess a store of knowledge about and rules for manipulating symbolic information. The goal of the researcher is to describe this collection of knowledge and rules at a useful level of detail. Computer languages are convenient formalisms for such descriptions because they were created for just this purpose—to describe how another symbol processing system, the computer, should manipulate symbols. Consequently, information-processing models are written in terms of computer programs.

Note that the claim is not that people think like computers, but that computers can be programmed to process information in a human-like way, at least at a desired level of detail. The computer program, not the computer itself, is the model of cognition. Implementing the model on a computer is usually necessary due to the complex interaction of its components. This complexity is required to capture the complexity of human information processing; however, it makes tracing the model's performance by hand difficult, if not impossible.

The computer is merely a tool for performing this difficult computation, in that same way that it is useful for doing statistical analyses of numerical data or calculating outcomes of other predictive, quantitative models like those of economics.

Most computer models try to capture human behavior at only one particular level of processing. Lower level processes are usually finessed and consequently, are probably not at all similar to human processes. For example, Jill Larkin and I have created a theory of the knowledge underlying the solving of simple arithmetic word problems such as "John has 8 marbles. He loses some of them. Now he has 5 marbles. How many marbles did he lose?" This theory is defined precisely by a series of computer models that simulate individuals with different degrees of skill in solving these problems. The models describe knowledge and rules children have for representing problems and for deciding how to manipulate objects (counters) to solve them. To solve the example problem, the model would make a set of eight counters to represent the eight marbles, then remove counters from this set until only 5 remained, and finally count the removed counters to determine the answer. Although the model explicitly describes knowledge underlying the representation of the problem and the use of that particular solution action, it does not specify knowledge underlying the necessary subskills such as counting. It always establishes the correct word-object correspondence and never makes mistakes like children do. One could, of course, create a model of knowledge underlying counting skill; in fact, Greeno, Riley and Gelman (Note 2) have done just that. Again, though, this model captures the processes of interest, establishing one-to-one correspondence between words and objects, but does not attempt to model the subskill of how number words are stored in memory, and how they are recalled. This model always says the number string flawlessly. Thus, most information-processing models attempt to describe processing only at a single level of detail, with no attempt to describe lower level processes or subskills in a realistic manner.

Currently, most computer models are written in programming languages called production systems. A production system has two parts. *Working memory* contains the data the system is considering at the moment and is analogous to human STM. *Production memory* is a collection of rules for manipulating information in working memory and is roughly analogous to human LTM. The rules in production memory are written as condition-action pairs called productions. The condition side of the production specifies one or more elements that must be in working memory for that production's actions to be executed (i.e., for it to fire). The action side specifies one or more actions to be done on elements in working memory—usually creating, deleting, or modifying elements. When all of the conditions of a production are matched by elements in working memory, the production's actions are carried out. Working memory then contains a new collection of elements which are subsequently compared to the condition sides of productions in memory.* If all the conditions of a production match elements in working memory, then its actions are executed, which again changes the elements in working memory. This *recognize-act* cycle continues until the set of elements in working memory do not satisfy the conditions of any production.

Production system programs differ from common types of programs like BASIC and FORTRAN in several ways. First, they can react flexibly to a variety of input. One can think of each production in memory as waiting to fire as soon as its conditions are satisfied. Thus, productions do not fire in a

---

*More than one production can have all of its conditions satisfied at the same time. In that case, the dominant production is determined according to predetermined conflict-resolution rules, which consider features such as specificity (the most specific production dominates) or recency (the production matching the most recent element in working memory dominates, keeping the system "on task"). See McDermott and Forgy, 1978, or Forgy, Note 3, for extensive discussions of conflict-resolution rules.

predetermined sequence. Further, the same production, or piece of knowledge, can be part of a variety of action sequences.

Second, production system programs are much easier to modify than FORTRAN or BASIC programs. Changing programs in these latter languages usually means rewriting a substantial portion of the programs and many times, it is easier simply to create a new program "from scratch." In contrast, production systems can be modified by merely adding or removing productions from memory. Models of different levels of skill or cognitive development thus can be constructed by adding, modifying, or removing sets of productions in memory. This does not imply that knowledge acquisition is simply quantitative, that it is nothing more than the incremental addition of skills. Because productions interact in complex ways, the addition of a new production or set of productions can drastically change both which productions fire and the sequence in which they fire. New productions alter the role of some old productions. Some no longer fire, while others can now be applied to new situations because one of the new productions produced working memory elements upon which the old production can act. As a result, the addition of several new productions can produce a qualitative, as well as quantitative, change in performance.

The ease with which production systems can be modified allows programs to be designed to modify themselves. This type of program contains productions whose actions build new productions in memory, rather than modify elements in working memory. These *adaptive production systems* are just beginning to be developed (e.g., Neves, Note 4; Neves and Anderson, 1981) but even at this early stage, they promise to represent a major contribution to the study of human learning. Learning is now commonly described by a series of computer models, each of which represents one level of skill acquisition, each providing only a "snapshot" view of learning. Adaptive production systems should allow us to go beyond these "snapshots" and create models that capture the learning process.

Computer models are evaluated by comparing their performance to that of humans. Any number of different parameters of performance can be compared, with the choice depending on the nature of the computer model and the task being modeled. Common parameters are error patterns in problem solving, reaction times, and sequences of steps in problem solving. Measures of the first two parameters are fairly straightforward. Determining the sequence of steps underlying a solution attempt is more complex, since much of the processing is covert, even in solving problems that involve manipulation of physical objects such as the Tower of Hanoi puzzle. Because information is thought to be processed in STM, what is really desired is data on the contents of STM and how it changes as a subject performs a task. Two ways of tracing the contents of STM have been developed: recording the sequence of a subject's eye-movements, and obtaining verbal protocols. The rationale underlying the eye-movement measure is that where a person is looking indicates the data being attended to. Consequently, the sequence of eye-movements indicates the sequence in which information was attended to, and hence, processed. A similar rationale underlies the verbal protocol data obtained from subjects "thinking aloud" while they solve a problem. Here, asking a person to "say whatever he is thinking" as he attempts to solve a problem is, in essence, asking him to report the contents of STM. Thus, verbal protocols provide at least a partial trace of the contents of STM during problem solving.* These sequences can be compared to the sequence in which the computer model processes information. The match between the two indicates the degree to which the computer program describes human information processing.

Creating explicit computer models of theories is important for several reasons. First, creating a computer program forces one to be precise, much in the way that writing an article forces one to be precise about one's ideas. Second, computer models provide tests of both the sufficiency of the knowledge specified in a theory to produce the desired behavior, and the internal consistency of the theory's assumptions (Greeno, 1980). Third, computer models clearly specify the consequences of a theory, even the unanticipated ones!

Computer modeling has its dangers, though. One is the strong temptation to stop modeling when a model capturing human behavior has been created. Just because a computer model is sufficient, consistent, and a reasonable simulation as determined by comparison of some parameters of performance, it may not be unique (i.e., necessary). Alternative models may be created that fit the data just as well. In this case, as in any situation of competing explanations, the question is whether the models differ in a substantive way, and if so, what data can distinguish between them. It should also be noted that it is often so difficult to create even one model that adequately simulates behavior, that alternative models seem a moot consideration.

Implications for Research on Mathematics Learning and Problem Solving

Information-processing methodology certainly has great potential benefit for the study of mathematics learning and problem solving. Some aspects, particularly the use of verbal protocols as data, are used increasingly in studies of mathematical problem solving (e.g., Kantowski, 1977; Schoenfeld, this volume). However, one of the major ideas from information-processing methodology—the creation of explicit models of complex human behavior— has yet to have much impact on research on mathematics learning. Explicit theories of mathematics learning and problem solving are long overdue. In the past, characterizations of mathematics learning have been variants of the theories of Piaget, Bruner, Dienes, Ausubel, and Polya, among others. Although these theories have contributed a number of important ideas, many concepts (e.g., stage of development, conservation, advance organ-

---

*See Ericsson and Simon, 1980, for an extensive discussion of verbal protocol as data.

izer, heuristic) are certainly not very precise in either definition or in their implications for mathematics instruction. It could be argued that it is not possible to characterize precisely some aspects of mathematics learning and problem solving, and thus it is impossible to create explicit models that capture the richness of the phenomena. However, much of this apparent inability is probably due to a shortage of detailed data about the phenomena (i.e., protocol or strategy data), and to limited technology, rather than to any inherent limitations stemming from the approach. For example, explicit models have been developed for complex phenomena such as discovering "new" mathematical concepts and conjectures (Lenat, Note 5). Thus, it seems that the creation of explicit models of mathematics learning and problem solving may be a fruitful direction for future research.

Advances in Characterizing Human Problem Solving

Initial information-processing research on human problem solving was concerned primarily with puzzle problems and games, like cryptarithmetic, the Tower of Hanoi, and missionaries and cannibals problems (Newell and Simon, 1972). This early work in "toy" domains enabled researchers to make important progress towards representing the problem-solving process (e.g., as a search in a problem space) and elucidating some general although weak methods (e.g., means-end analysis, hill climbing) that seemed to characterize human problem solving in unfamiliar domains.

Recent research, however, concerns problem solving in complex subject matter domains like physics, mathematics, computer programming and electronics. Although a comprehensive review of this newer research is beyond the scope of this paper, selected research will be described briefly to indicate the current direction of information-processing research.

One of the major new results is that successful problem solving in real, or "semantically rich" domains, depends on large amounts of domain-specific knowledge (Simon, 1979b). Studies contrasting performance of expert and novice problem solvers within a domain have found that experts use powerful, domain-specific strategies to solve problems, while novices rely on the weaker, more general methods, such as means-end analysis, that were identified in earlier research (Larkin, McDermott, Simon, and Simon, 1980; Simon and Simon, 1978; Chase and Simon, 1979.)

Studies of problem solving in semantically rich domains have also lead to a shift in the characterization of problem representations. In research on problem solving in toy domains, problem representations were described as state spaces or problem spaces, consisting of elements representing each state of knowledge about a task, operators, an initial knowledge state, a goal state, and additional knowledge available to the solver (Newell and Simon, 1972). Problem solving was described then as a search through this space. Recent studies of problem solving in subject matter domains have found it more useful to describe problem representations in terms of conceptually rich structures (Larkin, Note 6) or "running mental models" (de Kleer and Brown, 1981), with much less emphasis on state space characterizations. These studies also suggest successful and unsuccessful problem solvers differ in the problem representations they create. This notion seems particularly promising for characterizing individual differences in mathematical problem-solving skill.

Larkin (Note 6) provides one desription of these differences. In her study of expert and novice physics problem solvers, she has found it useful to distinguish among three types of problem representation. Naive representations center around entities that are familiar in everyday life, e.g., toboggans, coffee cups. Scientific representations center around entities that have special technical meaning that people learn ordinarily only through study of a science. Examples from physics include forces, momenta, and energy states. Mathematical representations are sets of equations reflecting physical principles applied to problems. Expert problem solvers used these rich scientific representations to guide their problem solving, while novice subjects relied on more naive representations.

Larkin and I (Note 1) have defined analogous distinctions among naive, mathematical, and computational representations in mathematics to describe differences in skill in mathematical problem solving. Here, naive representations also involve entities corresponding to items that might be directly perceived in the environment and inferencing rules that correspond to rules observed in real situations. They reflect understanding of everyday situations, but not their mathematical structure. Mathematical representations involve mathematical entities or concepts and the relations among them; the associated inferencing rules are time-independent and involve relations that are difficult to abstract from the environment without formal instruction. Finally, computational representations involve representing the problem using mathematical symbols along with inferencing rules about using computational algorithms. These distinctions have been useful for describing differences in skill in solving elementary word problems. In addition, they may be general enough to describe differences in problem-solving skill over a range of topics.

In recent years, a number of explicit computer models have been created to describe knowledge underlying a variety of mathematical concepts and skills, including: counting (Greeno, Riley, and Gelman, 1979), systematic errors in subtraction (Van Lehn, 1981), solving elementary word problems (Riley, Greeno, and Heller, in press; Briars and Larkin, Note 1), geometry (Greeno, 1978; Anderson, Greeno, Kline, and Neves, 1981), and learning the subtraction algorithm (Resnick, in press). Although it is impossible to characterize these models here, one comment is in order. The value of these models is not that they identify new overt phenomena or behaviors. In fact, most of them do not. For example, at a "common knowledge" level, if not a research level, it has been known that children make systematic errors in executing the subtraction (and other) algorithms, that planning is important in constructing geometry proofs, and that instruction in algorithms using concrete materials should make explicit the relation between the concrete and symbolic representations. However, the real value of these research studies is that they provide a detailed analysis of the knowledge and processes underlying these behaviors and describe them in terms of explicit models. This is not to say that these are the "correct" models, nor the only possible ones that could have been created. However, attempting to describe processes and knowledge involved in mathematical problem solving at such a level of detail may be a potentially rich direction for future research.

Summary

This paper has discussed some of the major concepts and constructs of information processing psychology and their implications for research on mathematics learning and problem solving. Specifically, the fundamental concept of people as information-processing systems and the resulting methodology, including the construction of computer simulation models, were described. The primary implications from this work are: first, the identification of variables and constructs (e.g., limited STM capacity, organization of knowledge in memory, and distinctions between procedural and declarative knowledge) that may be important to consider in future research on mathematics learning and problem solving, and second, the information-processing methodology, particularly the development of explicit models of behavior. Some recent advances in characterizing human problem solving were also described. This is not meant to suggest that no progress is being made in research on mathematical problem solving outside of an information processing perspective. Nor is it meant to claim that all information-processing research in general, or that on mathematical problem solving in particular, should profoundly influence research on mathematics learning and problem solving. What is being suggested, however, is that mathematical problem solving is just one activity of a complex organism, and that it could be studied more effectively by considering it from a more global perspective, in terms of what is known about cognition and human behavior, particularly from an information-processing perspective.

Reference Notes

1. Briars, D. J. and Larkin, J. H. *An Integrated Model of Skill in Solving Elementary Word Problems.* Paper presented at the Biennial Meeting of the Society for Research in Child Development, Boston, April, 1981.
2. Greeno, J. G., Riley, M. S., and Gelman, R. *Young Children's Counting and Understanding of Principles.* University of Pittsburgh, Learning Research and Development Center, unpublished manuscript, 1979.
3. Forgy, C. *The OPS5 Reference Manual.* Technical Report. Computer Science Department, Carnegie-Mellon University, 1981.
4. Neves, D. M. *Learning Procedures from Examples.* Unpublished Ph.D. thesis, Carnegie-Mellon University, 1981.
5. Lenat, D. B. *Proceedings of the Fifth International Joint Conference on Artificial Intelligence.* 1977, 833-842.
6. Larkin, J. H. *The Role of Problem Representation in Physics.* C.I.P.429, Carnegie-Mellon University, 1981. (Presented at a Conference on Mental Models, LaJolla, CA, 1980).

References

Anderson, J. R. *Language, Memory, and Thought.* Hillsdale, NJ: Lawrence Erlbaum Associates Inc., 1976.

Anderson, J. R. *Cognitive Psychology and Its Implications.* San Francisco: W. H. Freeman and Co., 1980.

Anderson, J. R., Greeno, J. G., Kline, P. L. and Neves, D. M. "Learning to Plan in Geometry." *In* J. R. Anderson (Ed.). *Cognitive Skills and Their Acquisition.* Hillsdale, NJ: Lawrence Erlbaum Associates Inc., 1981.

Case, R. "Piaget and Beyond: Toward a Developmentally Based Theory and Technology of Instruction." *In* R. Glaser (Ed.), *Advances in Instructional Psychology.* Hillsdale, NJ: Lawrence Erlbaum Associates Inc., 1978.

Chase, W. G. "Elementary Information Processes." *In* W. K. Estes (Ed.), *Handbook of Learning and Cognitive Processes, Volume 5: Human Information Processing.* Hillsdale, NJ: Lawrence Erlbaum Associates Inc., 1978.

Chase, W. G. and Ericsson, K. A. "Skilled Memory." *In* J. R. Anderson (Ed.), *Cognitive Skills and Their Acquisitions.* Hillsdale, NJ: Lawrence Erlbaum Associates Inc., 1981.

Chase, W. G. and Simon, H. A. "Perception in Chess." *In* H. A. Simon (Ed.), *Models of Thought.* New Haven, CT: Yale University Press, 1979.

Crowder, R. G. *Principles of Learning and Memory.* Hillsdale NJ: Lawrence Erlbaum Associates Inc., 1976.

de Kleer, J. and Brown, J. S. "Mental Models of Physical Mechanisms and Their Acquisition." *In* J. R. Anderson, (Ed.), *Cognitive Skills and Their Acquisition.* Hillsdale, NJ: Lawrence Erlbaum Associates Inc., 1981.

deGroot, A. D. *Thought and Choice in Chess.* The Hague: Mouton, 1965.

D. J. Briars

Ericsson, K. A. and Simon, H. A. "Verbal Reports as Data." *Psychological Review,* 1980, *87,* 215-251.

Estes, W. K. "The Information-processing Approach to Cognition: A Confluence of Metaphors and Methods." *In* W. K. Estes (Ed.), *Handbook of Learning and Cognitive Processes, Volume 5: Human Information Processing.* Hillsdale, NJ: Lawrence Erlbaum Associates Inc., 1978.

Feigenbaum, E. A. "The Simulation of Verbal Learning Behavior." *In Proceedings of the 1961 Western Joint Computer Conference.* 1961. Reprinted in E. A. Feigenbaum and J. Feldman (Eds.), *Computers and Thought.* New York: McGraw-Hill, 1963, pp. 297-309.

Fuson, K. C., Richards, J. and Briars, D. J. "The Acquisition and Elaboration of the Number Word Sequence." *In* C. J. Brainerd (Ed.), *Children's Logical and Mathematical Cognition.* New York: Springer-Verlag, in press.

Greeno, J. G. "A Study of Problem Solving." *In* R. Glaser (Ed.), *Advances in Instructional Psychology.* Hillsdale, NJ: Lawrence Erlbaum Associates Inc., 1978.

Greeno, J. G. "Psychology of Learning, 1960-1980: One Participant's Observations." *American Psychologist,* 1980, *35,* 713-728.

Hintzman, D. L. "Explorations With a Discrimination Net Model for Paired-associate Learning." *Journal of Mathematical Psychology,* 1968, *5,* 123-162.

Jeffries, R., Turner, A. A., Polson, P. G., and Atwood, M. E. "The Processes Involved in Designing Software." *In* J. R. Anderson (Ed.), *Cognitive Skills and Their Acquisition.* Hillsdale, NJ: Lawrence Erlbaum Associates Inc., 1981.

Kantowski, M. G. "Processes Involved in Mathematical Problem Solving." *Journal for Research in Mathematics Education,* 1977, *8,* 163-180.

Larkin, J. G., McDermott, J., Simon, D. P., and Simon, H. A. "Models of Competence in Solving Physics Problems." *Cognitive Science,* 1980, *4,* 317-345.

Lewis, C. "Skill in Algebra." *In* J. R. Anderson (Ed.), *Cognitive Skills and Their Acquisition.* Hillsdale, NJ: Lawrence Erlbaum Associates Inc., 1981.

McDermott, J. and Forgy, C. "Production System Conflict Resolution Strategies." *In* D. Waterman and F. Hayes-Roth (Eds.), *Pattern Directed Inference Systems.* New York: Academic Press, 1978.

Miller, G. A. "The Magical Number Seven, Plus or Minus Two: Some Limitations on Our Capacity for Processing Information." *Psychological Review.* 1956, *63,* 81-97.

Neves, D. M. and Anderson, J. R. "Becoming an Expert at Cognitive Skill." *In* J. R. Anderson (Ed.), *Cognitive Skills and Their Acquisition.* Hillsdale, NJ: Lawrence Erlbaum Associates Inc., 1981.

Newell, A. and Simon, H. A. *Human Problem Solving.* Englewood Cliffs, NJ: Prentice-Hall, Inc., 1972.

Resnick, L. B. "A Developmental Theory of Number Understanding." *In* H. P. Ginsburg (Ed.). *The Development of Mathematical Thinking.* New

York: Academic Press, in press.

Riley, M. S., Greeno, J. G., and Heller, J. L. "Development of Children's Problem Solving Ability in Arithmetic." *In* H. P. Ginsburg (Ed.), *The Development of Mathematical Thinking.* New York: Academic Press, in press.

Romberg, T. A. and Collis, K. F. "Cognitive Functioning and Performance on Addition and Subtraction Tasks." *In* T. R. Post and M. P. Roberts (Eds.), *Proceedings of the Third Annual Meeting of the North American Chapter of the International Group for the Psychology of Mathematics Education.* Minneapolis, MN, 1981, 137-141.

Rumelhart, D. E. "Notes on a Schema for Stories." *In* D. G. Bobrow, and A. Collins (Eds.), *Representation and Understanding.* New York: Academic Press, 1975.

Schank, R. and Abelson, R. P. *Scripts, Plans, Goals, and Understanding.* Hillsdale, NJ: Lawrence Erlbaum Associates Inc., 1977.

Silver, E. A. "Recall of Mathematical Problem Information: Solving Related Problems." *Journal for Research in Mathematics Education,* 1981. *12,* 54-64.

Simon, H. A. "Information-processing Theory of Human Problem Solving." *In* W. K. Estes (Ed.), *Handbook of Learning and Cognitive Processes, Volume 5: Human Information Processing.* Hillsdale, NJ: Lawrence Erlbaum Associates Inc., 1978.

Simon, H. A. "The Information-storage System Called 'Human Memory.'" *In* H. Simon (Ed.), *Models of Thought.* New Haven, CT: Yale University Press, 1979. (a)

Simon, H. A. "Information Processing Models of Cognition." *Annual Review of Psychology,* 1979, *30,* 363-396. (b)

Simon, H. A. "Studying Human Intelligence by Creating Artificial Intelligence." *American Scientist,* 1981, *69,* 300-309.

Simon, H. A. and Feigenbaum, E. A. "An Information Processing Theory of Some Effects of Similarity, Familiarity, and Meaningfulness in Verbal Learning." *Journal of Verbal Learning and Behavior,* 1964, *3,* 385-396.

Simon, H. A. and Gilmartin, K. "A Simulation of Memory for Chess Positions." *Cognitive Psychology,* 1973, *5,* 29-46.

Simon, D. P. and Simon, H. A. "Individual Differences in Solving Physics Problems." *In* Siegler, R. (Ed.), *Children's Thinking: What Develops?* Hillsdale, NJ: Lawrence Erlbaum Associates Inc., 1978.

Turvey, M. T. "Visual Processing and Short-term Memory." *In* W. K. Estes (Ed.), *Handbook of Learning and Cognitive Processes, Volume 5: Human Information Processing.* Hillsdale, NJ: Lawrence Erlbaum Associates Inc., 1978.

VanLehn, K. "On the Representation of Procedures in Repair Theory." *In* H. P. Ginsburg (Ed.), *The Development of Mathematical Thinking.* New York: Academic Press, in press.

Building Bridges Between Psychological and Mathematics Education Research on Problem Solving*

Frank K. Lester, Jr.

Research in that elusive area of human behavior called "problem solving" has become increasingly popular in recent years in several fields, among them education, psychology, sociology, and artificial intelligence. Within the field of education, mathematics educators have been especially active. Indeed, in a review of recent mathematical problem-solving research, I found that mathematics educators have devoted more attention to problem-solving research than to any other single area of inquiry (Lester, 1980). As was expected, I found an extremely wide range in the quality, focus, and methodology of this body of research as well as a general lack of agreement on what problem solving involves. I attributed this rather chaotic state to three factors: 1. a neglect of theory to guide systematic inquiry; 2. the extreme complexity of the nature of problem solving; and 3. the rudimentary state of the research methodologies employed. In both my review and a more recently written paper (Lester, in press), I discussed several broad issues associated with mathematical problem solving that need attention if the state of the research is to become more systematic and focused. Briefly, these issues relate to:

1. The role of theory in problem-solving research;
2. The types of research tasks to use;
3. The relative emphasis to place on developing competency models or performance models of problem-solving behavior;
4. Teaching problem solving (if in fact it can be taught) and what the teacher's role should be;
5. The nature of problem-solving performance changes;
6. The types of research methodologies to employ.

One step toward developing a proper perspective on these issues would be to look at how researchers in other fields have studied problem solving.

*This chapter is a portion of a larger working paper, "Psychological Problem Solving Research: Implications for Mathematics Education." I am indebted to Joe Garofalo and Sandy Kerr, both of Indiana University, Richard Mayer of the University of California-Santa Barbara, and Richard Shumway of The Ohio State University for their valuable suggestions for improving the working paper and consequently this manuscript. Partial research support for the preparation of this paper was provided by a grant from the Spencer Foundation.

Perhaps the most appropriate field to choose is cognitive psychology. Psychologists, both cognitive and other types, have a long history of looking to mathematics as a medium through which to study human learning and instruction. Indeed, some of the most prominent researchers in the history of psychology have been interested in mathematics learning—e.g., Thorndike, Brownell, Bruner, Gagné, Piaget, and Wertheimer. More recently, Greeno (1978), Mayer (1978), and Resnick (Resnick and Ford, 1981) have been interested in developing theories of mathematics learning in general, and mathematical problem solving in particular.

The purpose of this paper is to discuss some of the contemporary cognitive psychology research which is relevant to the aforementioned issues. The intent of this discussion is to move toward a resolution of these issues and to promote communication between psychologists and mathematics educators presently involved in problem-solving investigation.

Key Research Questions

The six issues posed in the preceding section are so broad that it would be unfeasible to try to survey the psychological research bearing on them. Therefore, I have identified seven questions, each related to one or more of the issues, around which to organize my discussion. They are:

1. Can problem solving be taught? (Related to issue 4).
2. What is the role of understanding in problem solving? (Issues 1 and 5).
3. To what extent does transfer of learning occur in problem solving? (Issue 5).
4. What are the primary task variables that affect problem solving? (Issues 1 and 2).
5. What is the role of metacognitive behavior in problem solving? (Issue 1).
6. How do successful and unsuccessful (good versus poor, expert versus novice) problem solvers differ with respect to their problem-solving behavior? (Issue 3).
7. What are the most appropriate research methodologies? (Issue 6).

For most mathematics educators, Question 1 may be the most important of all, since most of us have come to mathematics education research by way of mathematics teaching at some level. Questions 2 through 5 are natural adjuncts to Question 1. This is true because an interest in problem-solving instruction leads almost immediately to concern about the importance of understanding in problem solving and the extent to which learning to solve a certain type of problem helps in solving other types. At the same time, as any good teacher knows, a myriad of factors influence students' success. It is vital that the primary determinants of success be identified and incorporated into instruction if it is to be effective. Finally, any serious learner and teacher of mathematics recognizes that ability to solve problems involves much more than acquiring a collection of skills and techniques. Also, the ability to monitor progress during problem solving and an awareness of one's own capabilities and limitations are at least as important. Such abilities are decidedly metacognitive in nature.

Question 6 could be stated in an even more fundamental way: "Why is it that some people are good problem solvers and others are not?" Historically, educators have always been interested in individual differences. Thus, it is natural to expect that this interest would extend to problem solving.

Regarding Question 7, the complexity of the nature of problem solving, especially the fact that much of the most interesting behavior is covert, necessitates careful deliberation about how to go about observing, collecting, and interpreting the right kinds of data. Consequently, it appears that alternatives to the traditional experimental designs and techniques must be developed and used.

Clearly, it is not appropriate to attempt to consider each of these questions in any detail; such an examination would require an entire volume. Instead, the remainder of this chapter includes very brief treatments of certain questions (viz., Questions 1, 4, 5, and 6) and more detailed discussions of the others (viz., Questions 2, 3, and 7). The decision to focus my attention on Questions 2, 3, and 7 stems from the particular interest I have in them and the substantial body of existing cognitive psychology literature which is related to them.

Research Related to Questions 1, 4, 5, and 6

Question 1: Can problem solving be taught? The interest among mathematics educators in problem-solving research has grown to a large extent from their own mathematics studies and their attempts to teach students how to "do real mathematics." The sense of satisfaction, indeed exhilaration, resulting from having solved a difficult problem and the perplexity caused by students' inability to solve any but the most routine problems have impelled many of us to investigate the causes of these phenomena. Thus, it is natural to expect mathematics educators to view as fundamental those questions concerned with the extent to which a person can be taught to be a better problem solver and the conditions that most enhance success. And while progress is being made in these areas, the literature on mathematical problem-solving instruction is still largely based on folklore and the sage advice of master teachers like George Polya (1957, 1962, 1965). That is to say, it can be characterized as being grounded more in the intuition of good teachers (I have heard this intuition referred to as "heuristic rules of thumb") than in scientific experimentation.

Regrettably, cognitive psychology provides little help, except to suggest that we are asking the right questions (or at least the *same* questions). A big question for cognitive psychologists interested in problem-solving instruction revolves around the teachability of *general* problem solving. The dichotomy regarding the domain-independence of problem solving versus the domain-specificity of problem solving is central to most of their research.

Unfortunately, as Newell (1980) points out, the dichotomy is far from being resolved. The volume edited by Tuma and Reif (1980), especially the chapters by Greeno, Simon, Larkin, and Newell, offers extremely lucid discussions of the issues surrounding questions associated with problem-solving instruction. Despite such valuable expositions, neither mathematics education nor cognitive psychology has yet come up with a reasonable theory of problem-solving instruction, let alone any prescriptions for instruction which have broad application to mathematics classrooms.

Question 4: What are the primary task variables which affect problem solving? There is no doubt that the nature of the problem confronting an individual greatly influences performance. It is natural then to expect that, since problem solving is an important part of mathematics learning, there would be substantial interest among mathematics educators in determining the task variables that affect success. A particularly useful development in mathematics education has been the identification of categories of task variables (Goldin and McClintock, 1979). The four major categories are syntax, mathematical content and nonmathematical context, structure, and heuristic processes. There are at least three reasons why mathematics educators have wanted to identify important task variables. First, it helps the researcher understand how the task interacts with the total task environment (Kulm, 1979). Second, if the researcher is unaware of key properties of the research tasks used, or fails to describe them well enough for others to replicate them, the results will be of little value (Goldin, Note 1). Third, knowledge of the primary determinants of problem difficulty can aid teachers in matching problems to students' ability, level of experience or development, and other factors.

It is not surprising that cognitive psychologists have paid relatively little attention to task variables in their problem-solving investigations. Unlike most mathematics-education research which has used research tasks found in standard school mathematics curricula, psychological research has typically chosen or designed tasks to meet the needs of the research question at hand. Thus, the mathematics educator might ask: "Given problems of this type, what can I learn about human problem-solving behavior?" On the other hand, the cognitive psychologist might ask: "If I want to investigate a certain aspect of human problem-solving behavior, which tasks would be most appropriate?"*

In spite of this lack of interest in task variables research among cognitive psychologists, certain research results do have implications for future mathematical task variables research. In particular, studies by Hayes and Simon (1976, 1977) indicate that structure variables clearly are not the only task variables which influence performance. For example, Hayes and

---

*There are some notable exceptions to this generalization. Paige and Simon (1966) have studied algebra problem solving, Greeno (1978) has investigated high school geometry students' proofs, and Resnick et al. have done extensive analyses of students' performances on computational tasks (Resnick and Ford, 1981).

Simon's work suggests that certain changes in the form of the text of a problem that do not change its meaning can alter the problem's difficulty significantly (nonmathematical content variable changes according to Goldin and McClintock's classification).

Question 5: What is the role of metacognitive behavior in problem solving? A perusal of recent research on memory and cognitive development indicates that considerable attention is being paid to phenomena referred to as metacognitive processes. Briefly, metacognitive processes are those associated with knowledge and cognition about cognitive phenomena, or as Flavell puts it: " 'Metacognition' refers to one's knowledge concerning one's own cognitive processes and products or anything related to them, e.g., the learning-relevant properties of information or data" (Flavell, 1976, p. 232). Research in this area, while growing in interest for psychological researchers, has been the exclusive domain of developmental psychologists (e.g., Flavell, 1976; Flavell and Wellman, 1977), reading specialists (e.g., Brown, 1977, 1978; Myers and Paris, 1978), and special education researchers (e.g., Brown and Barclay, 1976; Meichenbaum and Asarnow, 1979).

I chose to include a question on metacognition because it is my belief that successful problem solving in mathematics is a function of at least five components: 1. mathematical knowledge and experience, 2. skill in the use of a variety of generic "tool" skills (e.g., sorting relevant from irrelevant information, drawing diagrams, etc.) 3. the ability to use a variety of heuristics known to be useful in mathematical problem solving, 4. knowledge about one's own cognitions before, during, and after a problem-solving episode, and 5. the ability to maintain executive control (i.e., to monitor and regulate) of the procedures being employed during problem solving. Thus, since metacognitive behaviors are involved in two of these five components, it is important that more attention be given to them. To date, mathematics educators and cognitive psychologists interested in problem solving have largely ignored metacognitive behaviors, the exception being the work of Silver, Branca, and Adams (1980). This is due primarily to a belief that metacognition is either unimportant or unresearchable.

Question 6: How do successful and unsuccessful problem solvers differ with respect to their problem-solving behavior? The study of the behavior of "experienced," "skilled," "expert," "successful," or "good" problem solvers has been popular in recent cognitive psychological research (e.g., Baron, 1978; Chase and Simon, 1973; deGroot, 1965; Krutetskii, 1976; and Larkin, 1980). This interest may stem at least in part from the fact that many cognitive psychologists have adopted an information-processing paradigm to guide their research. Information-processing psychologists are especially interested in skilful problem solving because of their interest in developing computer models of behavior; it is easier to model exemplary behavior than unskilled, unsuccessful, or poor behavior. In addition, artificial-intelligence researchers, who often address the same questions as cognitive psychologists, have been particularly concerned with creating systems that perform tasks well, not like "ordinary" humans (Schank, 1980; Bundy, Note 2). Consequently, they too have focused on "skilled" behavior.

Another approach has been to compare and contrast the problem-solving behaviors of expert and novice (or skilled and unskilled) subjects. This approach is illustrated by the research of Chi, Feltovich and Glaser (Note 3), Larkin, McDermott, Simon and Simon (1980), and Simon and Simon (1978).

While these two approaches may have appeal for mathematics education researchers, they have several inherent limitations. Perhaps the most serious limitation of studying only skilful behavior is that at least some of the processes used by skilful problem solvers may be qualitatively different from those used by less-skilful individuals. It may not be possible for unskilled problem solvers to acquire some of these processes due to such factors as developmental differences, cognitive style differences, or lack of experience in mathematics. Also, the fact that good problem solvers behave in certain ways does not imply that novice or poor problem solvers can be taught to behave in those ways. Finally, it must be kept in mind that behaviors that distinguish good from poor problem solvers are probably not sufficient for successful proving solving. Thus, even if novices can be taught the processes used by experts, it does not follow that these novices will become experts.

Research Related to Questions 2, 3, and 7

Question 2: What is the role of understanding in problem solving? Consider the following classroom scenario: Students are working diligently at their desks on a math assignment given them by their teacher. One student, who had been out of school with the flu, goes up to the teacher's desk and complains: "I don't know what to do! I don't even know how to start." The teacher, a firm believer in "discovery learning," tells him he can figure it out and instructs him to go back to his desk and think about it ("Use your head" is the teacher's advice). The boy obediently returns to his seat and stares blindly at the problem, becoming increasingly anxious and frustrated. He wonders why no one will tell him the secret to "how to think." (adapted from Brown, Collins, and Harris, 1977).

This scenario, an all-too-frequent occurrence in our classrooms, raises the question: Where are students being taught *how* to understand something new on their own? Moreover, where are they being taught what it means to understand?

The view that understanding is an important ingredient in all types of learning is so widely held within the mathematics education community that it appears to be regarded as an axiom of mathematics learning and instruction. The high esteem in which Brownell's "meaning theory" is held illustrates the support for this view (Brownell, 1935). Furthermore, the pre-eminence of Polya's four-stage problem-solving model, with "understanding the problem" as stage one, has caused both curriculum writers and mathematical problem-solving researchers to focus considerable attention on understanding. But although understanding has long been an aim of mathematics instruction, psychologists are beginning only now to attempt to provide a sound theoretical basis for this aim.* Realizing the importance of understanding has led to investigations of the ways in which understanding influences problem-solving performance.

It is appropriate to begin a discussion of the role of understanding by describing what it means to contemporary cognitive psychologists. Hayes and Simon (1976) provide an excellent illustration of the complex nature of understanding in their discussion of the Four Color Problem. This famous problem can be stated as follows:

> For any subdivision of the plane into non-overlapping territories, is it possible to color the plane with not more than four colors in such a way that no territories with a common border have the same color?

It is not difficult to imagine that a young child could grasp what the problem is about and what must be done to solve it (at least to the extent that the child understands at more than a syntactic level only). At the same time, an astute adult with no special training in mathematics might well recognize the need to do more than color a few maps. Also, a bright mathematics student might even realize that there are so many map configurations and ways to color the maps that a proof by exhaustion would not be feasible. Finally, it seems apparent that a research mathematician who had worked on this problem for years would have a different level of understanding of the problem than any of the others. Thus, there are degrees of understanding a problem, some of which come from the problem solver's background and some of which come from attempting to solve it, to say nothing of differences attributable to intellectual capacity, motivation, etc. Indeed, Simon (1975) suggests that understanding is impossible to define because it does not denote a single set of cognitive processes.

For cognitive psychologists, understanding is closely linked to the subject's internal representation of the problem. In particular, Greeno (1977, 1978) sees understanding as a process of constructing a representation of the object that is to be understood. He suggests that the difference between understanding and *not* understanding rests with the nature of the subject's representation. He also notes that since there is no unique representation for any problem, it is impossible to know whether a problem has been understood completely. Greeno's view is very similar to that of Schank (1972) and Winograd (1972), who have worked on the development of theories of understanding in natural language. For them, if a sentence is understood, the subject's internal representation shows what that sentence means. Greeno points out that this meaning ". . . corresponds to a pattern of relations among concepts that are mentioned in the sentence, and under-

---

*It should be mentioned that cognitive psychology has not so much *neglected* understanding as made it trivial by the assumptions of their theories of problem solving.

standing is the act of constructing such a pattern" (Greeno, 1977, p. 44). In reference to Hayes and Simon's discussion of levels of understanding of the Four Color Problem, the child's internal representation would be very different from the research mathematician's, although both might be accurate. The mathematician's representation would likely contain a more complex pattern of relations among the relevant elements of the problem.

Hayes and Simon (1977) provide a related description of what understanding involves. "To understand a written problem text, a person must do two things. First, he must read the sentences of the text and extract information from them by grammatical and semantic analysis. Second, he must construct from the newly extracted information a representation of the problem that is adequate for its solution. This representation must include the initial conditions of the problem, its goal, and the operators for reaching the goal from the initial state" (Hayes and Simon, 1977, p. 21). In a similar vein, Brown, Collins and Harris (1978) offer an artificial-intelligence description of understanding. For them, understanding a textual material requires the subject to construct an interpretation of the problem situation. This interpretation results from having three different kinds of knowledge which are not explicitly stated in the problem: 1, basic world (domain) knowledge relevant for the problem; 2, planning knowledge; and 3, strategic knowledge. Basic world knowledge typically is background, experience-based knowledge which must be used in order to make sense out of the problem. Planning knowledge enables the subject to "get a feel" for what needs to be done. It helps the individual understand what sequence of actions might lead to a solution. Strategic knowledge governs how basic world knowledge and planning knowledge are to be used in synthesizing a structural model of the meaning of the problem. Strategic knowledge, then, is metaknowledge that the individual uses to direct the processes of making sense out of a problem and deciding what actions to take to get a solution.

Computer models of human problem solving developed by Simon and his associates have employed two complex processes: an *understanding* process and a *solving* process. The understanding process contains two subprocesses: one for interpreting the language of the instructions, the other for constructing the problem space. After the problem space has been created, the solving process explores (searches) it to try to solve the problem. Thus, the solving process is totally dependent upon the understanding process. Furthermore, when a problem solver constructs a good internal representation (i.e., develops good understanding) of a problem, the solving process is simplified because the amount of searching needed to find a solution is reduced (Greeno, 1980).

The development of computer programs to direct machines to perform tasks associated with intelligent behavior has required artificial-intelligence researchers to develop certain axioms (formal assumptions about behavior) as a basis for analyzing cognitive processes, particularly processes associated with problem solving. These axioms are used to indicate structural and procedural mechanisms about problem solving and, more specifically, about ways to represent knowledge and understand textual material. Central questions for artificial-intelligence researchers are: "What does it mean to understand a text?" and "How does an individual come to understand a text?" (Brown, Collins, and Harris, 1977). Brown and his colleagues believe understanding processes are so important that problem-solving instruction should focus on them. A fundamental premise of their research is that the underlying domain-independent cognitive processes and knowledge that individuals must use to understand a situation, text, etc., should be explicated and ways should be found to teach students a general awareness of these processes and knowledge along with some learning strategies based on these processes. This way, students will have a foundation for acquiring new knowledge and their fear of new conceptual material they cannot instantly understand will decrease.* Among mathematics education researchers Charles (in press), Silver, Branca, and Adams (1980) and Stengel, LeBlanc, Jacobson and Lester (1977) espouse approaches to problem-solving instruction that are consistent with the premise put forth by Brown and his colleagues.

Greeno (1977) also agrees that the understanding phase is so important that it should be stressed in instruction. He believes the most effective understanding instruction would focus on problem structure and research support for this position is cited in a review by Mayer (1975). Based on the results of a series of experiments, Mayer concluded that when students were given "meaningful" instruction on a topic (e.g., solving algebra equations and binomial probability problems) they were better able to solve story problems, answer questions, and identify impossible solutions than were subjects whose instruction did not stress meaning. (Mayer describes a meaningful context as one which provides ". . . a means for subjects to relate the presented information to experiences and knowledge already in memory" [Mayer, 1978, p. 254].)

It is one thing to have general agreement about what a phenomenon is and what it involves and quite another thing to decide how to measure the presence of that phenomenon. Throughout most of this century psychologists (Gestaltists excepted) have been reluctant to address questions related to covert behavior. It is not surprising then that understanding has been particularly intractable, since it cannot be described by a single set of cognitive processes. It is easier to study a phenomenon, especially a covert one like understanding, if one can establish criteria for determining when and to what extent the phenomenon has occurred. Greeno (1977) has specified three criteria for good understanding: *1. achievement of a coherent representation; 2. close correspondence* between the internal representation and the object to be understood; *3. connectedness* of the representation to general concepts and procedures in a structure of knowledge. This is to say that a problem solver who has formed a representation which relates

---

*It is interesting to note the similarities between views like this and the tenets of meaningful learning long held by mathematics educators.

problem components in a compact structure understands the problem better than a person whose representation is poorly integrated (Criterion 1). At the same time, a person whose representation matches all the relevant components of a problem has better understanding than a person who has ignored or misinterpreted some of the important information, thereby creating a discrepancy between representation and the object understood (Criterion 2). Finally, good understanding has taken place when the understood object and its components are related to the problem solver's other knowledge. The more general the knowledge relating to the problem components that a person has, the better will be that person's understanding of the problem (Criterion 3).

Using these three criteria for understanding, Greeno investigated the performance of subjects on geometry problems. Thinking-aloud protocols were recorded from interviews held with high-school geometry students over an entire academic year in order to identify the kinds of representations they would construct as they worked on problems. A schematic diagram of each protocol was constructed which showed the components and relation that a subject was thinking about. (Greeno was quick to point out that these diagrams are only approximations of what the subject was actually thinking.) Greeno was able to judge the problem solver's degree of understanding by considering the properties of the solution pattern as indicated by the diagrams. In this way he was able to determine the extent to which the three criteria were satisfied and also to develop some hypotheses about what it means to solve a problem with understanding. These hypotheses led to the creation of a hypothetical problem-solving system called PERDIX. For Greeno, a primary theoretical aspiration for PERDIX was ". . . that the knowledge structures it needs to solve problems should be reasonable facsimiles of the kinds of concepts and principles that are taught in classroom instruction" (Greeno, 1977, p. 58). The main processes PERDIX uses are pattern matching and pattern generation, and, in this sense at least, it is not a very powerful system. However, it can satisfy the first of the three criteria of good understanding (viz., achieving a coherent representation).

To summarize, Greeno studied understanding in problems using the standard information-processing research paradigm: Collect and analyze problem solvers' protocols, construct a hypothetical system (computer model) based on these protocols, and test the system against predetermined criteria. A primary feature of his efforts was his almost exclusive concern with the manner by which subjects develop an understanding of problems rather than with correctness of solution or solution strategies. In addition, his use of schematic diagrams to analyze subjects' protocols seems very efficient and much less cumbersome than the coding schemes commonly used by mathematics educators. This procedure could be an effective protocol analysis tool for many types of mathematical tasks.

Hayes and Simon (1974, 1977) studied how individuals develop initial understanding of a problem by presenting problem texts and analyzing the ways problem solvers processed the texts. Subjects were given problems having the same formal structure as the Tower of Hanoi problem, the essential difference being in the context of the problem statements (e.g., "discs and pegs" were replaced by "monsters and globes"). Like Greeno, they analyzed problem-solving protocols in order to create a program (called UNDERSTAND) that could produce a problem space from a given problem text using a combination of syntactic and semantic relations. Hayes and Simon conceded that this computer simulation of human problem solving was far from perfect, but it matched very well the gross structure of the processes by which humans come to understand problem texts.

The analysis of the process of understanding written problems conducted by Hayes and Simon regarded the process as a translation from text to a set of problem-solving operators in a problem space. Recent analyses of physics problem solving have stressed understanding processes that form abstract structures of information that mediate between a problem text and problem-solving procedures (Larkin, McDermott, Simon, and Simon, 1980; Simon and Simon 1978). The basis for this different approach is that in understanding the information present in a problem, the problem solver grasps relationships (chunks) that allow the identification of subgoals and approaches to problem solving by relatively direct retrieval rather than by elaborate search as was true of the earlier model proposed by Hayes and Simon (Greeno, 1980).

I have tried to describe how contemporary cognitive psychologists view that part of the problem-solving process called "understanding." The view I have attempted to develop is that understanding processes are extremely complex and influence all other aspects of problem-solving behavior (planning, carrying out plans, evaluating solutions and answers, etc.). Moreover, many psychologists believe that the understanding stage is so important that problem-solving instruction is most effective when the development of understanding processes are stressed. Finally, I have shown that, despite the complexity and covert nature of understanding processes, they are no longer regarded as being inaccessible to systematic investigation.

Simon (1978) insists that psychological experiments are ignoring a vital part of the act of problem solving when they delay data gathering on problem-solving behavior until after the subject has had a chance to practice a task. In other words, problem-solving researchers cannot afford to restrict their attention to planning and solving behavior. There is too much else of importance that is also involved.

The message in the foregoing discussion is not limited to pointing out the importance of studying the role of understanding in problem solving. The message may be elucidated in the following points:

- Replications and extensions using mathematics problems of the best psychological research are sorely needed.
- The kind of understanding that an individual attains in a task domain can be important for ability to transfer acquired knowledge to similar tasks.

- Much current mathematics instruction gives insufficient attention to direct training in how to gain understanding of a problem.
- Understanding processes include the processes of developing representations of problem situations. Most problems can be represented in several ways.
- The problem representation a problem solver chooses may greatly affect problem difficulty.

Question 3: To what extent does transfer of learning occur in problem solving? A substantial amount of cognitive psychological research on problem solving can be considered transfer of learning research. Until recently much, if not most, of this research could be classified as transfer of *training,* that is, transfer following variations in training or instruction. (Such training typically involved practice in solving a problem or class of problems. No direct instruction or guidance was provided.) In his early review, Duncan (1959) observed that the results of research in this area were not clear cut. For example, it appeared that results were confounded by the interaction between method of training and the transfer tasks used. He called for research using complex problems (not just anagram and water jar problems) focusing attention on what specific responses are learned under a particular training method and on what responses are required on a particular transfer task. Finally, he saw a need for more emphasis on the amount and breadth of training. Among those studies which could be classified as transfer of learning experiments (i.e., experiments of the type described above in which experimental and control group performances are compared on a transfer task) *without* training or instruction, Duncan (1959) found changes in internal problem *structure* to be the only type of problem variation which consistently influences performance. Unfortunately, Duncan did not specify what the nature of this influence was.

Davis (1966) claimed that much of the transfer-related research involved "insight" problems (classified by Davis as tasks eliciting primarily covert trial-and-error behavior) and it is impossible to draw any general conclusions from the research for this reason alone. Duncan (1961) and Campbell (1960) found no evidence of any improvement of problem-solving performance by any pre-training methods they employed. In particular, in no case was there any nonspecific transfer from solving one task to another. They interpreted their results as supporting the view that solving insight problems is largely a trial-and-error-search. Another interesting result is found in a study by Hoffman, Burke, and Maier (1963). They found that prior experience with a simple version of Maier's hatrack problem led to more incorrect solutions on the standard hatrack problem than no prior experience at all: pre-training produced negative transfer. They explained this by suggesting that experimental subjects had acquired greater variability in their solution attempts and consequently more wrong solutions.

The notions of "functional fixedness" and "set" or *Einstellung* are commonly associated with research on problem-solving transfer. "Functional fixedness" is a term applied to large negative transfer produced by previous experience with certain stimuli (see Duncker, 1945; Maier, 1930). Saugstad and Rascheim (1960) attempted to show that functional fixedness is a direct result of the novel *functions* of objects being unavailable or unknown to the problem solver. In their experiment, in order to solve a problem, subjects had to bend a nail and use it as a hook, and roll up newspapers into tubes through which balls could pass. Before being given the problems, subjects were handed bent nails and rolled newspapers and asked to give examples of ways in which they might be used. As one would expect, subjects who did this solved the problems more readily than those who did not. Saugstad and Rascheim concluded that subjects who had available the necessary functions solved the problem. (I think they had actually reduced the problem to an absurdly simple task, not a problem at all.) In effect what they did was to tell subjects indirectly what actions were needed to solve the problem. This limitation notwithstanding, the authors claimed this experiment demonstrated that if the necessary functions are ' taught to subjects in advance, they will solve the problem without fixedness. Scheerer (1963) argued that inability to solve a problem is often due to the functionally fixed object not being available to the problem solver as an effective stimulus (i.e., the problem solver does not notice the object as being part of the solution materials).

Maier and Burke (1966) offered a different reason for the causes of functional fixedness. They surmised that successful and unsuccessful problem solvers differ, not in terms of the availability of relevant functions, but rather in terms of their selection of the proper functions to solve the problem.

The term "set" refers to a relatively fixed and rigid approach to problem solving brought about by pre-training. A commonly used task in "set" research is the water-jar problem popularized by Luchins (1942) in his classic and still-continuing series of experiments. Subjects solved a series of problems having the same solution (e.g., fill jar Y, pour from it into jar X until it is full, and into jar Z twice). A test problem required the subject either to solve: 1. an optional-solution problem where this method or an alternative, easier method can be used, or 2. a problem using a different method. The series of training problems made it likely that the subject would choose the old method with the optional task and would have difficulty with the task requiring a new method. The pre-training established a set in the subjects, a tendency to respond in one particular way rather than in other possible ways. Luchin's experiments illustrate set as inducing negative transfer. One interpretation of Luchins' results is that if subjects learn a particular solution to a problem they tend to stick to it for other problems even though it is not appropriate or not the most efficient. Less set is established if subjects are shown alternative solutions to the same problem. This interpretation, if valid, could have profound implications for classroom problem-solving instruction.

In my mind both set and functional fixedness (actually a special case of set) are phenomena associated with problem-solving behavior; they are not really explanations of behavior. Thus, while the experiments of Maier, Scheerer, Luchins, and others are very interesting, they seem to provide little information which can be used to explain *why* problem solvers behave the way they do. They do have instructional implications, however. For example, Luchins' research suggests that extensive training in the use of a particular technique or procedure may result in a set being established. The current "back-to-basics" movement, with its undue emphasis on drill, ignores the potential for harmful side effects, like establishing set.

Any discussion of transfer of learning in problem solving is complicated by the fact that the relationship between transfer of learning and problem solving has not been universally agreed upon. Schulz (1963) attempted to clarify this relationship by suggesting that problem solving be considered from a transfer of training point of view. With this view a "situation's status as a problem is largely contingent upon the fact that the sequence of performances conformed to the paradigm for negative transfer" (p. 175). In other words, a task becomes a problem when the problem solver has previously learned a response which is incompatible with that required for solution. In a similar vein, Gagné (1966) stressed the importance of transfer of learning in problem solving with his statement: "The most important characteristic of this newly acquired capability (i.e., an ability to solve a problem) is its lack of specificity, or to say it another way, its inherent generalizability. The investigator is simply not convinced that problem solving has occurred unless he performs what is in effect a transfer experiment" (p. 130). Gagné reserved the label "problem solving" for behavior which leads to the acquisition of a new principle. Furthermore, evidence that a new principle has been acquired must come from the problem solver's ability to solve a different, but related problem. On the other hand, Newell (1966) considered transfer to be a very inappropriate criterion for problem solving. He insisted that for some problems giving a solution is proof enough that problem solving has occurred, that there is no need to generalize or to show transfer to some other task. Greeno (1977) attempted to resolve the controversy by suggesting that, although it may be appropriate to consider problem solving to have occurred whenever a correct solution is obtained, a problem is solved with "good understanding" only when the problem solver recognizes the relation of the solution to some general principle. That is to say, true transfer of learning takes place only when the problem solver realizes that task *B* has certain structural properties in common with the previously-solved task *A*.

Until now no mention has been made of what is transferred from solving one problem to another. Recently some very interesting and thoughtful work has been conducted on this question by several researchers, most notably Reed (Reed, 1977, 1978; Reed, Ernst, and Banerji, 1974) and Hayes and Simon (1977). Their efforts also point out the limitations of much of the early transfer of learning research. For example, the information-processing theory of problem solving proposed by Newell and Simon (1972) and further developed by Simon and Greeno among others (e.g., Simon, 1978; Greeno,

1977) indicates the factors which influence problem solving, and hence, transfer. By specifying the type of mental activity required of the problem solver, the theory explains why it is not enough to study transfer simply in terms of strengthening correct response alternatives or in terms of constructs such as set and functional fixedness. The theory describes the processes of extracting structural information, encoding information (i.e., organizing input), and retrieving information. In addition, the theory details the role of perceptual processes, planning processes, and other types of cognitive activity. For this reason, contemporary information-processing theory provides the researcher with a means for determining what is transferred from one problem to another by identifying the key actions in problem solving and by suggesting the major sources of difficulty for the individual. Thus, the research is able to go beyond the question "Does transfer occur?" and ask, "What is transferred and under what conditions?"

Simon (1978) viewed "effective" problem solving as involving the extraction of information about the structure of the problem (task environment) and the use of that information for selective heuristic searches for a solution. Thus, it is not surprising to find that a great deal of research has investigated the effects of structural changes in tasks on problem-solving behavior. The research of Reed and his associates typifies this body of inquiry (Reed, Ernst, and Banerji, 1974; Reed and Johnsen, 1977; Reed, 1977). The purpose of his study with Ernst and Banerji was to investigate transfer between two homomorphic problems,* the missionaries and cannibals problem and the jealous husbands problem. In the first of three experiments, subjects solved both problems; half solved the missionary-cannibal problem first and half solved the jealous husbands problem first. The amount of transfer was measured by calculating the ratios $(MC_1 - MC_2)/MC_1$ and $(JH_1 - JH_2)/JH_1$, where MC_i, JH_i ($i = 1, 2$) denote solution times, MC_1 denotes the missionaries and cannibals problem was solved first, and MC_2, JH_1, and JH_2 are defined in a similar way. Subjects were not told the relationship between the two problems. Subjects in experiment two solved either the missionary-cannibal problem twice or the jealous husbands problem twice. Experiment three was the same as experiment one except that subjects were informed of the relationship between the two problems. Very little transfer occurred in experiment one. This suggested that the similarity of two task environments was not a sufficient condition for a large amount of transfer to occur. However, when subjects were told the relationship between the two problems (experiment 3) a significant amount of transfer took place, but only when the more complex problem (jealous husbands) was given first. In fact the amount of transfer was about the same as when the jealous husbands

---

*Problem homomorphism is used here to refer to state-space-homomorphism as defined by Goldin (1979). The reader is referred to Goldin's thorough discussion of state-space-homomorphism and in particular to his discussion of the homomorphisms between the missionaries and cannibals problem and the jealous husbands problem.

problem was solved twice. The results of the second experiment indicated that subjects showed nearly twice as much transfer in their second attempt at the jealous husbands problem than in their second attempt at the missionary-cannibal problem (i.e., $(JH_1 - JH_2)/JH_1 \approx 2 \cdot (MC_1 - MC_2)/MC_1$). Reed (1977) explained the results of experiment three by pointing out that the solution of the jealous husbands problem identified a unique solution for the missionary-cannibal problem but the reverse condition was not true. That is, the solution of the missionary-cannibal problem does not define a unique solution of the jealous husbands problem. Reed also cautioned against trying to generalize the results of his research. He pointed out that while the missionary-cannibal task and similar tasks have properties which make it attractive for research purposes, other problems may not possess these properties, consequently making it difficult to define the similarity between problems when transfer is being studied. He also recognized the limitations of transfer studies which are concerned only with measuring the amount of transfer. Such limited focus provides no possibility for identifying *what* is transferred (Reed, 1978). Is improved performance on a second problem due to the subjects having learned plans and general skills from problem one or to the fact they simply remembered a specific solution (or part of a solution)? Of course, the likely answer is that the better performance is due to both of these factors as well as others.

Whereas the study of Reed, Ernst, and Banerji (1974) looked at transfer between problem *homomorphs,* Hayes and Simon (1977) studied the extent of transfer from solving one problem to solving a different but *isomorphic* problem. In a series of three experiments they investigated the influence of changes in ". . . the form of the problem text on the representation of problems, and consequently upon the process of solution of problems, by humans" (p. 22). Their research was spurred by the notion that since changes in the representation of the problem dramatically alter the solution process, it is possible that changes in the problem statement might also affect the solution process. At the same time, by systematically varying the differences among isomorphic problems, the researcher can more easily study determinants of problem difficulty as well as transfer. Tasks used in the three experiments were eight variants to the Tower of Hanoi puzzle and involved "monsters" holding "globes" of three different sizes. The eight monster problems differed in three ways:

- Whether a *transfer of location* or a *change in size* by monsters or globes occurs in the problem (transfer versus change);
- Whether the monsters transfer or change *globes* or the monsters transfer or change *themselves* (agent versus patient); and
- The initial configuration of monsters and globes in the problem statement. The initial arrangement was either (SM, ML, LS) or (SL, MS, LM), where SM represents a small monster holding a medium-sized globe, and so on.

Results suggested that "change" problems were significantly more difficult than "transfer" problems (difficulty determined by solution time and number of failures to find solutions). In the first experiment, substantially more transfer occurred from easier to harder problems than the reverse condition. However, for "patient" problems an opposite but less marked result was found. The result for "agent" problems is counter to that of Reed, Ernst, and Banerji (1974), Dienes and Jeeves (1970), and Luger (1979). In the second experiment, a highly significant transfer of training effect was found. In contrast to the results of experiment one, transfer of training was always greater from the harder problem to the easier one than for the reverse. Also, it appeared that transfer of location versus change of size differences were more important to subjects' solution processes than agent versus patient differences. That is, variations in the transfer versus change dimensions had a much greater effect on subjects' solutions than did variations in the agent versus patient dimensions (it made relatively little difference if the problem was an agent or patient problem, but transfer problems were much easier for subjects than change problems).

A third experiment was performed to account for the differences in solution time between transfer and change problems. The results indicated that the solution time difference between these two types of problems can be attributed to differences in the difficulty of performing comparisons when applying the rules for a legal move. Hayes and Simon recognized that a complete analysis of transfer of training effects must include the identification of the processes that are facilitated by prior training. They suggested three such groups of processes:

- Processes for formulating the initial situation,
- Processes for identifying and representing the operations, and
- Processes for formulating the rules for a legal move.

Their analysis of subjects' solution efforts indicated that only the third group, processes for formulating the rules for a legal move, are likely to be the processes responsible for differences in transfer of training from a problem to one of its isomorphs.

To summarize, if the results of Hayes and Simon can be applied to other types of problems, the mere fact that two problems are structurally isomorphic does not imply they are of equal difficulty or that transfer will occur. Moreover, it appears that changes in the written instructions for a problem can have a profound effect on solution performance. They believe "this effect is produced because different problem instructions cause subjects to adopt different problem representations, even when the problems are formally isomorphic" (Hayes and Simon, 1977, p. 21). These results, of course, are far from surprising to anyone who has taught mathematics at any level. In fact, the staff of the Mathematical Problem Solving Project (MPSP) were clearly cognizant of the potential effect upon problem difficulty of changes in the problem statement. It is for this reason that "complexity of the problem statement" played such a prominent role in the problem categorization scheme devised by the MPSP (Lester, 1978). The value of

Hayes and Simon's research lies not in the finding that changes in text affect problem difficulty but rather in identifying *how* these changes influence problem representation and the solution processes used.

Acquiring a clearer understanding of the nature of transfer of learning should be of fundamental interest to all mathematics education researchers, particularly those involved in problem-solving research. A number of implications of the psychological research for mathematics education come to mind. Most obvious perhaps is the need for replications of the research of Reed, Simon, and Hayes, and others using mathematics problems. The research that has already been conducted by mathematics educators, notably by Kulm and Days (1979) and Silver (1981), have yielded results which differ in some respects from the results of the psychology studies. It may be that the nature of the problems used in mathematics instruction is so different from that of the "puzzle" problems used by Reed and by Hayes and Simon that it is not possible to apply their findings to mathematics problems. Before this conclusion is reached prematurely, more replications are in order.

Another suggestion is that research on the effect of instruction on transfer is needed. Several recent studies have looked at the effects of instruction on mathematical problem-solving behavior. Unfortunately, none of this research seems to have focused on transfer of training.

It also would be valuable to know more about what young children (e.g., primary grade students) transfer from solving one problem to a related problem and from solving one type of problem to another type. To my knowledge, there have been no transfer of learning problem-solving studies involving young children.

Finally, it would be appropriate to study the differences between good, average, and poor problem solvers with respect to transfer. The little research of this type using mathematics problems has given equivocal results (cf., Krutetskii, 1976; Silver, 1981). If mathematics teachers are to be expected both to provide good problem-solving instruction and to consider their students' individual differences, they must know how good, average, and poor problem solvers differ.

Question 5: What are the most appropriate research methodologies to employ? At least three questions come to mind in attempting to answer question five: These questions are: *1.* To what extent are qualitative methods appropriate? *2.* What types of research tasks should be used? and *3.* How should problem-solving performance be assessed? Each of these three questions is considered in the next several pages.

To what extent are qualitative methods appropriate? Cognitive psychologists, while recognizing the difficulties in attempting to identify and study covert behavior, believe that cognitive processes can and should be studied systematically. This stems from seeing problem solving as requiring the acquisition of cognitive structures quite distinct from other forms of learning. Also, they believe that almost everything of interest that goes on during problem solving is covert. Experimental, exclusively quantitative techniques therefore seem inadequate. In view of the current interest among mathematics educators in such nonexperimental methods as "teaching experiments" and protocol analysis, a discussion of psychologists' arguments for and against the use of qualitative methodologies is warranted.

The systematic description of cognitive phenomena using data based on verbal reports has a long history, dating back at least as far as the work of Wurzburg School psychologists at the turn of this century (Boring, 1950). DeGroot (1966), reporting on over 25 years of his own research, stated that introspection and protocol analysis have been used extensively by all researchers interested in process descriptions. More recently, Estes (1978) argues that sound theories of problem solving will result from the recording of factual accounts of what problem solvers do and from *inferred* concepts of covert behavior. In line with this position, Newell and Simon (1972) state that their research is empirically but not experimentally based.

Determining how to obtain data on problem-solving behavior is a particularly thorny problem concerning the use of qualitative procedures. The three most popular techniques are retrospection, "thinking aloud," and introspection by trained observers. The first two techniques require the problem solver to provide a verbal report of his or her behavior either after completion of a problem-solving episode (retrospection) or during problem solving (thinking aloud). The relative merits of these two approaches have been argued thoroughly in a series of recent articles appearing in the *Psychological Review* (Nisbett and Wilson, 1977; Smith and Miller, 1978; White, 1980; and Ericsson and Simon, 1980). This series was motivated at least in part by Nisbett and Wilson's thoughtfully prepared review of research which suggests that individuals have *no* direct access to higher order mental processes. They posited the proposition that when individuals try to report on their mental processes, they do not do so on the basis of true introspection. Rather, they report ". . . by applying or generating causal theories about the effects of that type of response. They simply make judgments, in other words, about how plausible it is that the stimulus would have influenced the response. These plausibility judgments exist prior to, or at least independently of, any actual contact with the particular stimulus embedded in a particular complex configuration" (Nisbett and Wilson, 1977, p. 248).

A quite valuable part of Nisbett and Wilson's paper is their discussion of the conditions under which verbal reports are likely to be accurate or inaccurate. They suggest that reports are accurate when influential stimuli are available to the problem solver and are viewed by him as plausible causes of the response, and when plausible but noninfluential factors are available. More importantly, they cite six sets of factors under which verbal reports will be inaccurate. These are:

1. A separation in time between the report and occurrence of the process (i.e., the subject tends to forget as time elapses);
2. Ignorance of certain influential factors associated with "the mechanics of judgement" (e.g., serial-order effects, position effects, contrast

effects, and anchoring effects);
3. Failure to regard contextual factors as salient;
4. Basing judgments and evaluations on noninfluential events which have occurred rather than on influential nonevents;
5. Failure to regard nonverbal behavior as important to evaluations relative to verbal behaviors; and
6. Failure to recognize that small causes can produce large effects. (I hasten to add that Nisbett and Wilson's conclusions were based to a large extent on results of research involving retrospective behavior).

If Nisbett and Wilson's arguments are accepted, verbal reports, especially retrospective ones, have little to recommend them as a source of information on the mental processes employed during problem solving. However, three retorts to their paper have appeared, suggesting that their theoretical stance is not well formulated, that some of the experiments they cite to support their case have serious design problems, and that the inaccurate verbal reports found by their research resulted from requesting information that was not directly heeded by the subjects (Smith and Miller, 1978; Ericsson and Simon, 1980; White, 1980).

To elaborate a bit on the criticisms made by the above-mentioned papers, Smith and Miller (1978) and White (1980) agree that the length of time between process and report can be a serious problem. However, they insist that many of the studies discussed by Nisbett and Wilson involved an inordinate time lag between process and report. White states: "It is a strain on cognitive capacity to keep in mind the route by which the solution to a problem was achieved after the achievement of that solution; in fact the task of remembering each stage is likely to act to the detriment of performance on the next stage" (p. 106). Furthermore, White argues that in addition to finding it difficult to remember a process, a problem solver may not think it worth the bother *unless* he/she is informed beforehand of the necessity of doing so. Unfortunately, the subjects in Nisbett and Wilson's research were not so informed. It appears that White was arguing *for* "thinking aloud" as well as *against* the types of retrospective reports discussed by Nisbett and Wilson.

The paper by Ericsson and Simon (1980) was not written solely for the purpose of refuting Nisbett and Wilson's conclusion. Their primary aim was to explicate a model of how individuals, in response to instructions to think aloud, verbalize information they are attending to in short-term memory. The model predicts what can be reliably reported and it can be used to aid in the interpretation of verbal data and the relationship between a problem solver's verbal and other behaviors. The model is based on four assumptions:

1. A human is an information-processing system;
2. Information is stored in several memories which have different capacities and accessing characteristics (viz., several sensory stores of short duration; short-term memory [STM], with limited capacity and/or short duration; long-term memory [LTM], with large capacity

but relatively slow fixation or access times);

3. Information recently acquired by the central processor (CP) is kept in STM and is directly accessible for further processing;
4. Information from LTM must be retrieved before it can be reported.

Due to the limited capacity of STM, only the most recently heeded information is directly accessible. However, part of the contents of STM are fixated in LTM before being lost or replaced. This portion can *sometimes* be retrieved later from LTM. Thus, any verbal report of cognitive processes would necessarily be based on only a subset of the information in STM and LTM. For this reason, and because of the above-mentioned assumptions about the model, a taxonomy of verbalization procedures results. This taxonomy is illustrated in Table 1, which is reproduced directly from Ericsson and Simon's paper.

Table 1. A Classification of Different Types of Verbalization Procedures as a Function of Time of Verbalization (Rows) and the Mapping From Heeded to Verbalized Information (Columns)

Time of verbalization	Relation between heeded and verbalized information			
	Direct one to one	Intermediate processing		
		Many to one	Unclear	No Relation
While information is attended	Talk aloud Think aloud	Intermediate inference and generative processes		
While information is still in short-term memory	Concurrent probing			
After the completion of the task-directed processes	Retrospective probing	Requests for general reports	Probing hypothetical states	Probing general states

Ericsson, K. A. and Simon, H. A. "Verbal Reports as Data." *Psychological Review*, 1980, 87 (3), page 224. Copyright 1980 by the American Psychological Association. Reprinted by permission of the author.

The table indicates that two dimensions are responsible for the types of verbalization procedures. First, the time of verbalization is important in deciding from what type of memory the information will be drawn. The second dimension distinguishes between procedures in which verbalization is a direct manifestation of stored information (direct, one-to-one) and procedures in which stored information is input to intermediate processes (e.g., abstraction, inference).

Their criticism of Nisbett and Wilson's paper is presented with respect to their model and I will not discuss these criticisms except to mention that they see a primary reason for the inaccurate verbal reports cited by Nisbett and Wilson as stemming from the fact that subjects were not asked to remember their mental processes.

Ericsson and Simon's article is an important contribution to problem solving research and should be read carefully by every mathematics educator contemplating the use of verbal reports of problem solving. In addition to their carefully articulated model of how problem solvers verbalize their thoughts, they make a number of points which deserve mention:

- Verbalizing information during problem solving (i.e., "thinking aloud") affects cognitive processes only if the information which must be verbalized would not otherwise be attended to.
- There is little published literature on methodological issues associated with verbal reports. Data-gathering and data-analysis methods vary tremendously and the details of these methods are only sketchily reported in research publications. This is a totally unsatisfactory state of affairs.
- No clear guidelines have been provided to distinguish between acceptable and unacceptable introspection.
- No distinction is made in the research literature between such diverse forms of verbalization as "thinking aloud," retrospective reports (both free recall and responses to specific probes), and classical introspective reports of trained observers.
- The failure of problem solvers to report some information does not invalidate verbal reports.
- If verbal reports are to be used as legitimate data in psychological research five issues must be considered: 1. the effects on the cognitive processes of the instruction to verbalize and the effects of probes; 2. the completeness of verbal reports; 3. the consistency of the verbal reports with other empirical data; 4. the generalizability and validity of the verbal reports; and 5. the design of objective methods of encoding and analyzing thinking-aloud protocols.

What does the foregoing discussion suggest for mathematical problem solving research? Consider the following:

1. The use of verbal reports as a source of data has been an issue for quite a long time in the psychology literature. Recent developments, especially by Simon and his associates, suggest that verbal reports should be regarded as a legitimate form of data.
2. Don't throw out the baby with the bath water. That is, the growing popularity of verbal reports and other qualitative data should not mean that quantitative data and quantitative analysis should be discarded.
3. Current work being done to develop methods for analyzing verbal protocols (e.g., Lucas et al., 1979), while a slow and painstaking task, is essential and must continue.

4. In another paper (Lester, 1980), I noted Hatfield's (1978) comment that mathematics education researchers who have analyzed problem-solving protocols have been content to study the type and frequency of occurrence of certain processes. That is to say, they have analyzed protocols quantitatively only. This single-minded approach to data analysis must change if progress is to be made.
5. Researchers who use verbal protocols should report the procedures they used to gather and analyze their data in more detail. With this in mind, journals which publish such reports should insist upon the inclusion of clear descriptions of methods of data collection and analysis.

What types of research tasks should be used? A very important consideration in conceptualizing and designing problem-solving research is the choice of tasks to use. There is ample reason to believe that the tasks used affect generalizability of results as well as the types of questions and hypotheses which can be investigated. Each of the several reviews I have consulted has pointed out that the state-of-the-art of problem-solving research can be partially characterized by the extreme diversity of tasks (Duncan, 1959; Davis, 1966; Dodd and Bourne, 1973; Shulman and Elstein, 1975). It is safe to say that no classification of problem-solving tasks has been satisfactory.

Davis (1966) suggested that despite the wide variability among tasks used in the research, in most tasks the response alternatives are not clearly specified for the problem solver. Consequently, these tasks require the subject to test and reject response options; that is, the subject must act in a trial-and-error fashion. This view of problem solving as trial-and-error learning led Davis to identify two categories of problem-solving tasks. The two categories, overt-type problems and covert-type problems, are distinguished on the basis of "whether the problem solver can or cannot associate a particular outcome or function to each of the available response alternatives" (p. 41). Examples of covert problems are anagrams (no aids available), water-jar problems and "insight" problems. Covert tasks are usually concrete, involve mentalistic concepts (e.g., set), and are scored on all-or-none basis (unless time is used). With overt-type tasks the subject must manifestly test various response options to determine their potential outcomes. Representative of this type of task are classification tasks, switch-light problems (the subject must obtain a particular pattern of lights by operating switches), and probability learning tasks. Overt tasks typically are abstract and involve behavioristic concepts (e.g., S-R associations). Scoring is usually continuous and multidimensional. According to Davis, one virtue of this analysis of research tasks is that it points out some continuity between what has often seemed a rather heterogeneous set of situations. However, it seems to me that only the covert tasks lend themselves to the study of problem solving. Overt tasks, as he describes them, seem to involve concept learning more than problem solving. Also, I think this categorization characterizes the propensity of researchers to use tasks for certain purposes, but I do not think it describes anything inherent in the tasks.

Dodd and Bourne (1973) identified six types of problems as being the ones most commonly used. Their types, corresponding very closely to Davis' types, were: water-jar problems, mental arithmetic, anagrams, switch-light problems, probability learning, and "insight" problems. Dodd and Bourne claim that anagrams were the most frequently used type of task, an observation also made by Davis (1973).

A final comment regarding the types of research tasks found in the literature is that psychologists have selected tasks which they feel best enable them to study particular aspects of problem-solving behavior. This explains, at least in part, why so many different types of problems appear. It also points out an extremely important difference between psychological and mathematical problem-solving research. For the psychologists the task used is simply a means to investigate a question about human problem solving. Thus, the task can be manipulated to fit the question. On the other hand, the mathematics educator does not have this luxury—mathematical problems must be used or at least the implications for mathematics education of the research must be borne in mind. Students are expected to learn to solve story problems, for example. Consequently, story problems should be used as tasks in at least some of the research. We cannot rely on research results based on the use of artificial problems and expect to say with confidence anything about mathematical problem solving. As I have mentioned earlier, it is important that mathematics educators consider the work done in psychology and undertake replications with mathematical tasks. Also, we would do well to begin to conduct research on narrowly-defined questions using clearly-defined problems and to conduct several closely related studies using the same problems. This is the only way that a stable body of knowledge about mathematical problem solving can be developed.

How should problem solving performance be assessed? The increased importance of problem solving in school mathematics curricula has made this question a particularly important one for mathematics educators. There has been general dissatisfaction with existing standardized problem-solving tests which are scored solely on the basis of correctness of answers. This dissatisfaction has caused several mathematics educators (e.g., Proudfit, 1977; Romberg and Wearne, 1975; Vos, 1976; and Schoen and Oehmke, Note 4) to develop instruments which measure processes as well as other aspects of problem solving.

The relatively few psychological researchers who have attended to this matter have been educational psychologists who undertook test development in order to measure creativity or divergent thinking, or to evaluate an instructional program (e.g., Covington, et al., 1972; Feldhusen and Houtz, 1975; Torrance, 1966). Such tests have not been limited to paper-and-pencil varieties but have included simulations, class projects, and the like.

Conclusion

Mathematics education, as a distinctly separate field of inquiry within education, is a relatively young discipline. Indeed, it can be argued that mathematics education developed an identity of its own only 62 years ago with the founding of the National Council of Teachers of Mathematics in 1920. In the early years the most influential "mathematics educators" were professional mathematicians with an interest in teaching. For the most part, trends were established and direction was provided by individuals who had little, if any, formal training in pedagogy or psychology, and who made judgments largely on the basis of their own intuition and experiences as teachers. Thus, while school mathematics curricula improved steadily under the thoughtful leadership of these mathematicians-turned-educators, it is not surprising that there was little change in the extent of knowledge about how people learn mathematics and how mathematics should be taught.

Today's mathematics educators have received a very different kind of preparation. They have received reasonably strong training in mathematics and have taught in "real" classrooms. In addition they have studied human growth and development, learning and cognition, and educational research methodology. In short, contemporary mathematics educators are better equipped than ever before to resolve the research issues confronting them.

Research in mathematics education is, and probably always will be, a rather esoteric field of educational research due to the very nature of mathematics. However, not only should mathematics education researchers look to other fields for direction and perspective, but also they should endeavor to communicate the results of their investigations to others outside of mathematics education. This seems particularly appropriate with respect to problem-solving research, since problem solving has interest for researchers in so many disciplines. I hope this paper will serve to stimulate dialogue about problem solving and promote cooperative efforts between mathematics educators and cognitive psychologists. If this paper does not serve as a framework for a bridge between these two groups, perhaps it will at least point out the need for such a bridge.

Reference Notes

1. Goldin, G. A. "The Measurement of Problem Solving Outcomes." Paper prepared for the conference on Issues and Directions in Mathematical Problem-Solving Research. Indiana University, Bloomington, May, 1981.
2. Bundy, A. *Analysing Mathematical Proofs.* Research report #2. Department of Artificial Intelligence, University of Edinburgh, 1975.
3. Chi, M., Feltovich, P. J., and Glaser, R. *Representation of Physics Knowledge by Experts and Novices.* (Technical Report 2). Pittsburgh: University of Pittsburgh Learning Research and Development Center, 1980.
4. Schoen, H. L. and Oehmke, T. *Iowa Problem Solving Test.* Iowa City: University of Iowa, 1980.

References

Baron, J. "Intelligence and General Strategies." *In* G. Underwood (Ed.) *Strategies of Information Processing.* London: Academic Press, 1978, 403-450.

Boring, E. G. *A History of Experimental Psychology* (2nd ed.) New York: Appleton-Century-Crofts, 1950.

Brown, A. L. "Knowing When, Where, and How to Remember: A Problem of Metacognition." *In* R. Glaser (Ed.), *Advances in Instructional Psychology.* Hillsdale, NJ: Lawrence Erlbaum Associates Inc., 1978.

Brown, A. L. "Metacognitive Development in Reading." *In* R. J. Spiro, B. Bruce, and W. F. Brewer (Eds.), *Theoretical Issues in Reading Comprehension.* Hillsdale, NJ: Lawrence Erlbaum Associates Inc., 1977.

Brown, A. L. and Barclay, C. "The Effects of Training Specific Mnemonics on the Metamnemonic Efficacy of Retarded Children." *Child Development,* 1976, *47,* 71-80.

Brown, J. S., Collins, A., and Harris, G. "Artificial Intelligence and Learning Strategies" ERIC #154-775, report date June 1977 (also in H. O'Neill [Ed.] *Learning Strategies.* New York: Academic Press Inc., 1978.

Brownell, W. A. "Psychological Considerations in the Learning and the Teaching of Arithmetic." *In The Teaching of Arithmetic,* Tenth Yearbook of the National Council of Teachers of Mathematics. Washington, D. C.: The National Council of Teachers of Mathematics, 1935, 1-31.

Campbell, D. T. "Blind Variation and Selective Retention in Creative Thought as in Other Knowledge Processes." *Psychological Review,* 1960, *67,* 380-400.

Charles, R. I. "An Instructional System for Mathematical Problem Solving." *In* S. L. Rachlin (Ed.), *MCATA Monograph on Problem Solving.* Calgary, Canada: University of Calgary (in press).

Chase, W. G. and Simon, H. A. "Perception in Chess." *Cognitive Psychology,* 1973, *4,* 55-81.

Covington, M. V., Crutchfield, R. S., Davies, L., and Olton, R. M. *The Productive Thinking Program.* Columbus, Ohio: Charles E. Merrill Publishing Co., 1972.

Davis, G. A. *Psychology of Problem Solving: Theory and Practice.* New York: Basic Books Inc., Publishers, 1973.

Davis, G. A. "Current Status of Research and Theory in Human Problem Solving." *Psychological Bulletin,* 1966, *66* (1), 36-54.

de Groot, A. D. "Perception and Memory Versus Thinking." *In* B. Kleinmuntz (Ed.), *Problem Solving: Research, Method, and Theory.* New York: John Wiley & Sons Inc., 1966.

de Groot, A. D. *Thought and Choice in Chess.* The Hague: Mouton, 1965.

Dienes, Z. P. and Jeeves, M. A. *The Effects of Structural Relations on Transfer.* London: Hutchinson Educational, 1970.

Dodd, D. H. and Bourne, L. E. "Thinking and Problem Solving." *In* B. B. Wolman (Ed.), *Handbook of General Psychology.* Englewood Cliffs, NJ: Prentice-Hall Inc., 1973.

Duncan, C. P. "Recent Research on Human Problem Solving." *Psychological Bulletin,* 1959, *56,* 397-429.

Duncan, C. P. "Attempts to Influence Performance on an Insight Problem." *Psychological Reports,* 1961, *9,* 35-42.

Duncker, K. "On Problem Solving." *Psychological Monographs,* 1945, *58* (5, Whole No. 270).

Ericsson, K. A. and Simon, H. A. "Verbal Reports as Data." *Psychological Review,* 1980, *87* (3), 215-251.

Estes, W. K. "The Information-processing Approach to Cognition: A Confluence of Metaphors and Methods." *In* W. K. Estes (Ed.), *Handbook of Learning and Cognitive Processes: Vol. 5, Human Information Processing.* Hillsdale, NJ: Lawrence Erlbaum Associates Inc., 1978.

Feldhusen, J. and Houtz, J. "Problem Solving and the Concrete Abstract Dimension." *The Gifted Child Quarterly.* 1975, *19* (2), 122-129.

Flavell, J. "Metacognitive Aspects of Problem Solving." *In* L. B. Resnick (Ed.), *The Nature of Intelligence.* Hillsdale, NJ: Lawrence Erlbaum Associates Inc., 1976.

Flavell, J. and Wellman, H. "Metamemory." *In* R. Kail and J. Hagen (Eds.), *Perpsectives on the Development of Memory and Cognition.* Hillsdale, NJ: Lawrence Erlbaum Associates Inc., 1977.

Gagne, R. M. "Human Problem Solving: Internal and External Events." *In* B. Kleinmuntz (Ed.), *Problem-Solving: Research, Method, and Theory.* New York: John Wiley & Sons Inc., 1966, 128-148.

Goldin, G. A. "Structure Variables in Problem Solving." In G. Goldin and C. E. McClintock (Eds.), *Task Variables in Mathematical Problem Solving.* Columbus, OH: ERIC/SMEAC, 1979.

Goldin, G. A. and McClintock, C. E. (Eds.), *Task Variables in Mathematical Problem Solving.* Columbus, OH: ERIC/SMEAC, 1979.

Greeno, J. G. "Process of Understanding in Problem Solving." *In* Castellan, N.J., Pisoni, D. B., and Potts, G. R. (Eds.), *Cognitive Theory,* Vol. 2.

Hillsdale, NJ: Lawrence Erlbaum Associates Inc., 1977, 43-83.

Greeno, J. G. "Natures of Problem Solving Abilities." In W. K. Estes (Ed.), *Handbook of Learning and Cognitive Processes: Vol. 5, Human Information Processing.* Hillsdale, NJ: Lawrence Erlbaum Associates Inc., 1978, 239-270.

Greeno, J. G. "A Theory of Knowledge for Problem Solving." *In* D. T. Tuma and F. Reif (Eds.), *Problem Solving and Education: Issues in Teaching and Research.* Hillsdale, NJ: Lawrence Erlbaum Associates Inc., 1980.

Hatfield, L. L. "Heuristic Emphases in the Instruction of Mathematical Problem Solving: Rationales and Research." *In* L. L. Hatfield and D. A. Bradbard (Eds.), *Mathematical Problem Solving: Papers From a Research Workshop.* Columbus, Ohio: ERIC/SMEAC, 1978.

Hayes, J. R. and Simon, H. A. "Understanding Written Problem Instructions." *In* L. Gregg, (Ed.), *Knowledge and Cognition.* Hillsdale, NJ: Lawrence Erlbaum Associates Inc., 1974.

Hayes, J. R. and Simon, H. A. "The Understanding Process: Problem Isomorphs." *Cognitive Psychology,* 1976, *8,* 165-190.

Hayes, J. R. and Simon, H. A. "Psychological Differences Among Problem Isomorphs." *In* N.J., Castellan, D. B. Pisoni, and G. R. Potts (Eds.), *Cognitive Theory,* Vol. 2. Hillsdale, NJ: Lawrence Erlbaum Associates Inc., 1977, 21-41.

Hoffman, L. R., Burke, R. J., and Maier, N. R. F. "Does Training With Differential Reinforcement on Similar Problem Help in Solving a New Problem?" *Psychological Reports,* 1963, *13,* 147-154.

Krutetskii, V. A. *The Psychology of Mathematical Abilities in School-children.* Chicago: University of Chicago Press, 1976.

Kulm, G. and Days, H. "Information Transfer in Solving Problems." *Journal for Research in Mathematics Eduction,* 1979, *10* (2), 94-102.

Kulm, G. "The Classification of Problem Solving Research Variables." *In* G. A. Goldin and C. E. McClintock (Eds.), *Task Variables in Mathematical Problem Solving.* Columbus, OH: ERIC/SMEAC, 1979.

Larkin, J. H. "Skilled Problem Solving in Physics: A Hierarchical Planning Model." *Journal of Structural Learning,* 1980, *6,* 271-297.

Larkin, J., McDermott, Simon, D. P., and Simon, H. A. "Expert and Novice Performance in Solving Physics Problems." *Science,* 1980, *208,* 1335-1342.

Lester, F. K. "Mathematical Problem Solving in the Elementary School: Some Educational and Psychological Considerations." *In* L. L. Hatfield and D. A. Bradbard (Eds.), *Mathematical Problem Solving: Papers From a Research Workshop.* Columbus, OH: ERIC/SMEAC, 1978.

Lester, F. K. "Research in Mathematical Problem Solving." *In* R. Shumway (Ed.), *Research in Mathematics Education.* Reston, Virginia: National Council of Teachers of Mathematics, 1980.

Lester, F. K. "Trends and Issues in Problem-solving Research." *In* R. Lesh and M. Landau (Eds.), *Acquisition of Mathematics Concepts and Processes.* New York: Academic Press Inc. (in press).

Lucas, J. F., Branca, N., Goldberg, D., Kantowski, M. B., Kellogg, H., and Smith, J. P. "A Process-sequence Coding System for Behavioral Analysis of Mathematical Problem Solving." In G. Goldin and C. E. McClintock (Eds.), *Task Variables in Mathematical Problem Solving.* Columbus, OH: ERIC/SMEAC, 1979.

Luchins, A. S. "Mechanization in Problem Solving: The Effect of Einstellung." *Psychological Monograph,* 1942, *54* (#6, Whole #248).

Luger, G. F. "State-space Representation of Problem Solving Behavior." *In* G. Goldin and C. E. McClintock (Eds.), *Task Variables in Mathematical Problem Solving.* Columbus, Ohio: ERIC/SMEAC, 1979.

Maier, N. R. F. "Reasoning in Humans I: On Direction." *Journal of Comparative and Physiological Psychology,* 1930, *10,* 114-143.

Maier, N. R. F. and Burke, R. J. "Test of the Concept of Availability of Functions in Problem Solving." *Psychological Reports,* 1966, *19,* 119-125.

Mayer, R. E. "Information Processing Variables in Learning to Solve Problems." *Review of Educational Research,* 1975, *45,* 525-541.

Mayer, R. E. "Effects of Meaningfulness on the Representation of Knowledge and the Process of Inference for Mathematical Problem Solving." *In* R. Revlin and R. E. Mayer (Eds.), *Human Reasoning,* New York: John Wiley & Sons Inc., 1978, 207-241.

Meichenbaum, D. and Asarnow, J. "Cognitive Behavioral Modification and Metacognitive Development: Implications for the Classroom." *In Cognitive-behavioral Interventions: Theory, Research, and Procedures.* New York: Academic Press Inc., 1979.

Myers, M. and Paris, S. G. "Children's Metacognition About Reading." *Journal of Educational Psychology,* 1978, *70* (5), 680-690.

Newell, A. "Discussion of Papers by Dr. Gagne and Dr. Hayes." *In* B. Kleinmuntz (Ed.), *Problem Solving: Research, Method, and Theory.* New York: John Wiley & Sons Inc., 1966, 171-182.

Newell, A. "A Final Word." In F. Tuma and F. Reif (Eds.), *Problem Solving and Education: Issues in Teaching and Research.* Hillsdale, NJ: Lawrence Erlbaum Associates Inc., 1980.

Newell, A. and Simon, H. A. *Human Problem Solving.* Englewood Cliffs, NJ: Prentice-Hall Inc., 1972.

Nisbett, R. E. and Wilson, T. D. "Telling More Than We Can Know: Verbal Reports on Mental Processes." *Psychological Review,* 1977, *84* (3), 231-259.

Paige, J. M. and Simon, H. A. "Cognitive Processes in Solving Algebra Word Problems." *In* B. Kleinmuntz (Ed.), *Problem Solving: Research, Method, Theory.* NY: John Wiley & Sons Inc., 1966, 51-110.

Polya, G. *How to Solve It* (2nd ed.). NY: Doubleday & Co. Inc., 1957.

Polya, G. *Mathematical Discovery: On Understanding, Learning and Teaching Problem Solving* (2 volumes), NY: John Wiley & Sons Inc., 1962, 1965.

Proudfit, L. *The Development of a Process Evaluation Instrument* (Technical Report V). Mathematical Problem Solving Project. Bloomington, IN:

Mathematics Education Development Center, 1977.

Reed, S. K. "Facilitation of Problem Solving." In N.J., Castellan, D. B. Pisoni, and G. R. Potts, *Cognitive Theory*, Vol. 2. Hillsdale, NJ: Lawrence Erlbaum Associates Inc., 1977, 3-20.

Reed, S. K. "Individual Differences in Problem Solving: General Heuristics of Specific Plans." *In* J. Scandura and C. J. Brainerd (Eds.), *Structural/ Process Models of Complex Human Behavior.* Proceedings of NATO Advanced Study Institute on Human Behavior, 1978. 603-612.

Reed, S. K., Ernst, G. W., and Banerji, R. "The Role of Analogy in Transfer Between Similar Problem States." *Cognitive Psychology*, 1974, *6*, 436-450.

Reed, S. K. and Johnsen, J. A. "Memory for Problem Solutions." *In* G. H. Bower, (Ed.), *The Psychology of Learning and Motivation.* Vol. II. New York: Academic Press Inc., 1978.

Resnick, L. B. and Ford, W. W. *The Psychology of Mathematics for Instruction.* Hillsdale, NJ: Lawrence Erlbaum Associates Inc., 1981.

Romberg, T. A. and Wearne, D. *Romberg-Wearne Mathematics Problem-solving Test.* Madison, WI: Research and Development Center for Cognitive Learning. The University of Wisconsin, 1975.

Saugstad, P. and Rasheim, K. "Problem Solving, Past Experience and Availability of Functions." *British Journal of Psychology*, 1960, *51*, 91-104.

Schank, R. C. "Conceptual Dependency: A Theory of Natural Language Understanding." *Cognitive Psychology*, 1972, *3*, 552-631.

Schank, R. C. "How Much Intelligence is There in Artificial Intelligence?" *Intelligence*, 1980, *4*, 1-14.

Scheerer, M. "Problem Solving." *Scientific American*, 1963, *208* (#4), 118-128.

Schulz, R. W. "Problem Solving Behavior and Transfer." *In* R. R. Grose and R. C. Birney (Eds.) *Transfer of Learning.* Princeton, NJ: Van Nostrand Reinhold Co., 1963 (also in *Harvard Educational Review*, 1960, *30*, 61-77).

Shulman, L. S. and Elstein, A. S. "Studies of Problem Solving, Judgment, and Decision Making: Implications for Educational Research." *In* F. N. Kerlinger (Ed.). *Review of Research in Education* (Vol. 3, p. 3-42). Itasca, IL: F. E. Peacock Publishers, Inc., 1975.

Silver, E. A. "Recall of Mathematical Problem Information: Solving Related Problems." *Journal for Research in Mathematics Education*, 1981, *12* (1), 54-64.

Silver, E. A., Branca, N.A., and Adams, V. M. "Metacognition: The Missing Link in Problem Solving?" *Proceedings of the Fourth International Conference for the Psychology of Mathematics Education*, Berkeley, California, August, 1980, 213-221.

Simon, H. A. "Learning With Understanding." *Mathematics Education Information Reports*, Columbus, Ohio: ERIC/SMEAC, 1975.

Simon, H. A. "Information-processing Theory of Human Problem Solving." *In* W. K. Estes (Ed.). *Handbook of Learning and Cognitive Processes: Vol. 5, Human Information Processing.* Hillsdale, NJ: Lawrence Erlbaum Associates Inc., 1978.

Simon, D. P. and Simon, H. A. "Individual Differences in Solving Physics Problems." *In* R. Siegler (Ed.), *Children's Thinking: What Develops?* Hillsdale, NJ: Lawrence Erlbaum Associates Inc., 1978.

Smith, E. R. and Miller, F. S. "Limits on Perception of Cognitive Processes: A Reply to Nisbett and Wilson." *Psychological Review,* 1978, 85 (4), 355-362.

Stengel, A., LeBlanc, J., Jacobson, M., and Lester, F. "Learning to Solve Problems by Solving Problems" (Technical Report II, D). Bloomington, IN: Mathematical Problem Solving Project, Indiana University, 1977.

Torrance, E. P. *Torrance Tests of Creative Thinking.* Princeton, NJ: Princeton Press, 1966.

Tuma, F. and Reif, F. (Eds.). *Problem Solving and Education: Issues in Teaching and Research.* Hillsdale, NJ: Lawrence Erlbaum Associates Inc., 1980.

Vos, K. "The Effects of Three Instructional Strategies on Problem Solving Behavior in Secondary School Mathematics." *Journal for Research in Mathematics Education,* 1976, 7 (5), 264-275.

White, P. "Limitations on Verbal Reports of Internal Events: A Refutation of Nisbett and Wilson and of Bem." *Psychological Review,* 1980, 87 (1) 105-112.

Winograd, T. "Understanding Natural Language." *Cognitive Psychology,* 1972, 3, 1-191.

The Measure of Problem-solving Outcomes

Gerald A. Goldin

Introduction

Several alternate definitions of the terms "problem" and "problem solving" have been proposed in the psychological literature (Lester, Note 1). For example, we have the following: "A question for which there is at the moment no answer is a problem" (Skinner, 1966, p. 225); "A problem arises when a living creature has a goal but does not know how this goal is to be reached" (Duncker, 1945, p. 1); "(A problem is) any situation in which the end result cannot be reached immediately" (Radford and Burton, 1974, p. 39); "(A problem is) a stimulus situation for which an organism does not have a ready response" (Davis, 1973, p. 12); "A person is confronted with a *problem* when he wants something and does not know immediately what series of actions he can perform to get it" (Newell and Simon, 1972, p. 72). This paper will take the operational view that the definition of problem solving must include a description of how it is to be measured. Two essential ingredients of such measurement are first, the task that is posed, and second, the recording of features of the events which subsequently ensue.

Some of the above definitions appear to restrict the domain of tasks which are to be considered "problems." Thus it may seem that merely by choosing a task for experimental investigation, we are already implicitly adopting some sort of definition. However the various limitations on the choice of task that are proposed in the definitions make reference to observations made *after* the task has been posed. Clearly we cannot know in advance whether or not the individual has a ready response, or whether the end result can be reached immediately. Logically, then, it is within the class of behaviors which occur after the presentation of the task that we must distinguish between "problem-solving" and "non-problem-solving" responses. To me it seems unlikely that the immediacy of the response will turn out to be the only important defining criterion.

The next section of this paper offers some observations concerning the task as a measuring instrument for the study of mathematical problem solving. The following section discusses various choices that have been made as to which events that occur after a task is posed are to be recorded, and how they are to be recorded. The kinds of measurements which have been made are organized under the following headings: Measures of problem difficulty, error patterns, strategy scoring systems, structured interview responses, paths through external state-spaces, and, finally, verbal, written, or enactive protocols. The final section returns to the question of defining problem solving. It suggests that if researchers could agree on standardized methods of defining and recording problem-solving outcomes for a wide range of tasks, they would find it easier to compare their observations.

The Task as a Measuring Instrument in Research*

One of the ideas motivating recent studies of task variables has been that of improving the replicability of problem-solving experiments (Goldin and McClintock, 1979; Kulm, 1979). Despite widespread agreement on the importance of replicating empirical findings, sufficient information to repeat problem-solving studies is rarely published, and replicating seldom is performed. While difficulties are inevitable in defining comparable subject populations, or preparing identical experimental treatments, the problem tasks themselves are crucial in establishing a standard of replicability.

It is interesting to compare the measurement of problem solving with the measurement of a physical property of a system such as its temperature. To perform even a coarse measurement of temperature, several steps are necessary to develop and calibrate a thermometer. One must first make the crude observation that mercury rises in a thin tube when the reservoir is placed in water that feels hot to the touch. One next establishes that the final height of the mercury depends on certain physical characteristics of the thermometer, such as the reservoir volume and the bore width. Thus it would be erroneous, for example, to report results in centimeters, ignoring these instrument variables. Standard temperatures must be defined with respect to reproducible states (for example, boiling and freezing points of water), and one must observe that these standards can be affected by additional variables (such as the atmospheric pressure) which require specification. Finally, a scale is established by calibrating the thermometer linearly between the standard temperatures, a step which must be justified by further experiment.

Similarly, if the problem-solving researcher is unaware of important properties of the task itself, or does not describe it well enough for other researchers to construct identical instruments, the observations made will be of limited use. In some studies, a single problem or set of problems is presented to a number of subjects, and their problem-solving processes are described. This is perhaps analogous to placing the tube of mercury in hot water, and describing the rise in level to its final height. Without information on how the various characteristics of the problem task(s) are influencing the observed processes, the observations cannot be interpreted as measurements. It is not possible to separate knowledge gained about the task itself from knowledge gained about the problem solvers.

In addition, small changes in a problem's syntax, content, context, or

*This section of this paper is adapted from a presentation by the author at the Fourth International Congress on Mathematical Education, Berkeley, CA, August, 1980.

structure can cause large changes in problem difficulty and in the processes used by subjects. Unlike the thermometer's measurements, problem-solving outcomes do not change slowly and continuously with the instrument variables. Because of this sensitivity to task characteristics, interpreting experimental findings usually requires detailed knowledge of the tasks used. But published articles rarely provide task information sufficient for unambiguous interpretation.

Some studies examine performance on different but related tasks. If, as is often the case, the tasks vary in many characteristics simultaneously, outcome differences cannot be attributed to particular task variables, and interpretation of the findings is again subject to limitations. To make progress in developing instrumentation for the study of problem solving, it is important to control as many task variables as possible, changing one characteristic at a time to observe its effects. The best way to learn how to build a thermometer is to vary a single physical characteristic at a time.

Just as standard temperatures are defined by reproducible physical states, problem characteristics need to be defined by accessible "reference populations" of problem solvers. A problem only "has" a given syntax for a population of speakers of standard English. Similarly it "has" a given structure for a population which uses certain mathematical notations and rules of procedure. Nevertheless syntax and structure can be described independently of individual problem solvers within the population. The use of reference populations is standard procedure for psychometricians who carry out item analyses on multiple-choice tests, but the practice seems to have received little attention among those studying problem solving by other means.

In problem-solving experiments, as in the measurement of physical quantities, it is important to know in advance the set of possible outcomes. Then the meaning of a particular observation can be understood in relation to other outcomes which might have occurred but did not. Without this advance knowledge, we again cannot tell whether the experiment is providing information about the problem solvers, or merely additional unforeseen information about the tasks. Reference populations may be helpful in establishing appropriate quantitative and classificatory scales for reporting problem-solving outcomes. Methods of defining and recording outcomes are discussed further in the next section.

The above considerations suggest that an exhaustive understanding of the tasks used in problem-solving research is needed to study problem solving more scientifically. It would be worthwhile for researchers to agree on even a small set of standardized problems, for which a substantial body of information on task characteristics for various populations could be developed. A classification scheme for task variables has been proposed by Kulm (1979), and various authors have offered detailed descriptions of task variables within each category (Barnett, 1979; Webb, 1979; Goldin, 1979; McClintock, 1979). Thus the foundation for such an approach has been laid.

G. A. Goldin

Scoring Systems for Problem-solving Studies

After a task has been selected and posed to subjects in an experiment, there are a number of ways to describe what happens next. Some of these methods depend on the task that is selected and are called "task-specific." Other methods are "task-independent," in that their definition does not hinge on the choice of task. In between these extremes is the possibility that a scoring system is applicable to a category of tasks which have features in common.

Measures of Problem Difficulty

Given a population to which a particular task is posed under specified conditions, the most elementary outcome measure is the fraction f of the population correctly solving the task. This quantity may be looked at in two ways. When we think of the population held constant and the task varied, f may be considered as a task attribute (showing the ease of solution). Alternatively, when we think of the task as held constant and the population varied, as between experimental and control groups, f may be regarded as an attribute of the population (showing the degree of problem-solving success).

Another measure of problem difficulty is the mean or the median time to solution for members of the population, or, more generally, the distribution of solution times. Such a measure is useful when all or nearly all subjects correctly solve the problem presented.

As naive as they are, these measures are basic to problem-solving research. Perhaps their most frequent application is to gauge the effectiveness of instructional treatments in studies where the tasks are post-test items (and sometimes pretest items as well). The "time to solution" measure has been used in the study of learning transfer between related problems (Reed, Ernst and Banerji, 1974; Luger and Bauer, 1978; Luger, 1979; Waters, 1979a, b). Dienes and Jeeves (1970) proposed the "deep end" hypothesis, which states generally that learning to solve related tasks occurs most efficiently when the more difficult task is solved first. This is tested by comparing the times to solution of populations solving a pair of problems in opposite order.

Another application of problem difficulty measures is in determining the effects of task characteristics on various populations. For example Barnett (1979) reviews a class of studies (e.g. Suppes, Loftus, and Jerman, 1969; Jerman and Rees, 1972; Jerman and Mirman, 1974; Barnett, 1974) based on a linear regression model in which the quantity-log f was assumed to depend linearly on certain task variables (as f varies from 1 to 0, its negative logarithm varies from 0 to infinity). Caldwell and I compared the difficulties of verbal problems differing with respect to abstract-concrete and factual-hypothetical variables, using the fraction f as a measure of ease of solution (Caldwell, 1977; Caldwell and Goldin, 1979; Goldin and Caldwell, 1979). In the linear regression studies many task characteristics were varied simultaneously, while the studies with Caldwell offer a model for controlling all but the task variables of immediate experimental interest.

A variation of these difficulty measures is to view the problem parts, one part a prerequisite for the next, in accordance with some kind of task analysis. The measures can now be applied separately to each of the problem components. Alternatively, a "partial credit" score can be assigned each subject based on the successful completion of portions of the task. For these scores to be meaningful, most or all of the subjects must approach the task via the indicated subtasks. Thus, while the "fraction correct" and the "time to solution" are measures which are task-independent and directly comparable for different tasks administered to the same population, we begin now to enter the domain of task-specific measures. A partial scoring system will yield comparable scores for a set of problems only if the same method of task analysis can be applied uniformly to subdivide all problems in the set into subtasks which can be scored in parallel.

Error Patterns

The problem difficulty measures discussed thus far provide very limited information about problem solving. The goals of problem-solving research must include not only the prediction of problem difficulties when population, task, and situation variables are fixed, but also the influencing of problem difficulties through instructional treatment—that is, maximizing the decrease in difficulty across the broadest possible range of tasks, with minimum of instruction. Information which may bear on this goal can be gained through measures examining patterns of errors made by subjects. I shall discuss two ways in which this has been done.

One approach describes the patterns of correct and incorrect answers offered by a single subject over a set of related tasks. This might be called the "diagnostic testing" approach. Assuming that the tasks require the use of certain well-defined rules or skills, information about a subject's competence in those skills can be inferred from the observed error patterns (Durnin, 1971; Gramick, 1975; Goldin and Gramick, 1980). Clearly such a "diagnostic profile" outcome is specific to the domain of tasks. It is also specific to the population, in that successful subjects are to be classified as having actually used the specified rules or skills (and not some alternative method).

A second approach describes the actual erroneous solutions offered by subjects on a single task or set of related tasks, classifying these under "type of error" headings with respect to a task analysis. For example, Hershkowitz, Vinner, and Bruckheimer (1980) classify children's mistakes in the addition of fractions as pre-algorithmic, motivated by the idea of a common denominator but missing the idea of expanding the fractions, or embodying both ideas but achieving one or the other incorrectly. The categories are necessarily highly task-specific. A more general error hierarchy for verbal arithmetic problems has been developed by Newman (1977) and utilized by Clements (1980). Here the error categories are reading, comprehension, transformation, process skills, encoding, and carelessness or motivation. The more specific categories of Hershkowitz et al. can be interpreted as an elaboration of Newman's "process skills" category for one domain of tasks.

Strategy Scoring Systems

Several researchers have used the approach of defining several ideal solution strategies for the task. A single strategy, if perfectly followed, would lead to a sequence of steps within a well-defined class of possible behaviors. It is rare, however, for a subject to follow a single strategy from beginning to end. One therefore assigns a "strategy score" which is a quotient: the number of steps within the class associated with the strategy, divided by the total number of steps. The strategy score is interpreted as a measure of the degree the strategy was used by each subject, thus providing information beyond simple measures of success or error classifications.

In their study of concept acquisition tasks, Bruner, Goodnow, and Austin (1956) distinguished among four strategies; conservative focusing, focus-gambling, successive scanning, and simultaneous scanning. These strategies were initially identified through techniques or protocol analysis. Laughlin (1965) defined strategy scoring methods based on subjects' discrete behaviors, arriving at a "focusing index" which was a quotient of focusing choices and total choices. Laughlin was aware of difficulties with such indices. For example, as the ideal strategies were initially defined, certain choices could be interpreted as resulting from more than one strategy. These ambiguities were difficult to resolve. Waters (1979a, b) points out additional difficulties with these strategy scores. While high focusing scores were associated with greater problem-solving success, this was a *logical necessity* based on the way the set of focusing choices had been delineated, rather than a psychological observation.

Dienes and Jeeves (1965, 1970) also used strategy scores to study the learning of rules for predicting the appearances of cards in a window where the rules were in fact based on the underlying structure of a mathematical group. They defined "operator" scores and "pattern" scores based on the choices made by subjects, but again logical difficulties arose in delineating the classes of choices associated with each strategy (Branca and Kilpatrick, 1972; Goldin, 1979, pp. 111-113). These difficulties limit the value of the finding that high operator scores were associated with greatest success.

While the strategies defined for these tasks appear to be highly task-specific, there are some features of the strategies which can be compared *across* tasks. For example, the most effective strategies, "focusing" and "operator" strategies, have the common feature of holding certain elements of the task environment fixed while others are consciously varied. This describes the responses to Piagetian pendulum tasks which are classified as "formal operational."

If we consider for a moment the Embedded Figures Test used to classify subjects as "field-independent" or "field-dependent" in cognitive style (based on successfully locating the embedded figures), it may be reasoned that an effective strategy is to "focus" on one fairly unusual feature of the figure and then search for that feature in the more complicated design. Perhaps the greater success of field-independent subjects on certain tasks can be accounted for in terms of certain kinds of "focusing" strategy usage. In any case it is clear that if the logical difficulties associated with strategy scoring systems could be overcome, they could provide an important and replicable means of classifying certain behavioral problem-solving outcomes.

Structured Interview Responses

More light can be shed on strategy usage by presenting subjects with structured questions which can be asked either during or after the problem-solving episode. Of course, when questions are asked during the course of problem solving, it cannot be assumed that the problem solving is unaffected by the questioning.

Retrospective accounts of strategy usage were recorded by Dienes and Jeeves (1965, 1970). They found a positive relationship between subjects' retrospective evaluations and the observed (behavioral) strategy scores. On the other hand, Branca and Kilpatrick (1972) found that retrospective evaluations frequently did not correspond to their measured strategy scores.

Waters (1979a, b) took a different approach by defining "intended focusing" and "intended scanning" indices for Bruner-type concept acquisition tasks. These indices were based not on subjects' actual choices, but on the reasons given for each choice. By questioning the subjects after each step, and scoring with reference to a pre-established set of possible responses, some of the difficulties associated with behavioral strategy scores could be eliminated.

Structured interviews have been used widely to study children's mathematical concepts from a developmental viewpoint. The key idea is to obtain additional information concerning a child's cognitive structures, schemata, etc. These interviews allow for questions contingent on the particular observed behavior (for example, Bessot and Comiti, 1978). The structured interview makes it often possible not only to classify the strategies employed by subjects, but to observe directly the efficacy of these strategies. Of course, to ensure that response patterns can be meaningfully interpreted, the categories of responses to structured questions must be established in advance.

Paths Through External State-Spaces

Representations of tasks can be described by means of state-spaces (Nilsson, 1971). A state-space is a set of distinguishable task configurations, called states, together with the permitted steps from one state to another, called moves. A state-space thus has the structure of a directed graph. A particular state is designated as the initial state, and one or more states are designated as goal states. The behavior of a problem solver can be recorded by the sequence of states entered during problem solving, providing still another measure for the observer.

Some applications of such observations follow. *1.* The number of states entered can be a measure of problem difficulty (for example Thomas, 1974; Reed, Ernst, and Banerji, 1974; Luger and Bauer, 1978). This measure permits comparisons for distinct populations solving the same problem, or for populations solving distinct problems with isomorphic external state-spaces. *2.* The lengths of time spent in each state can be observed and compared, as can the proportions of incorrect or illegal moves originating from different states (Thomas, 1974). *3.* Moves can be organized into clusters by virtue of the pauses which occur between them (Greeno, 1974). *4.* Patterns in state-space paths can be identified and associated with the identification of subgoals, the occurrence of problem-solving stages, or insight into problem symmetry (Goldin, 1979; Luger, 1979).

While tasks which are not isomorphic will have different state-spaces, there are ways of describing relatedness between tasks by means of state-space homomorphisms (Goldin, 1979). This permits some comparison of the state-space paths generated by subjects who solve homomorphic problems. With this exception, state-space paths are a highly task-specific measure of problem solving, so much so that one cannot directly compare the paths of subjects solving the same problem by means of distinct representations.

Verbal, Written, and Enactive Protocols

This category has been reserved for studies which rely on subjects' "thinking aloud" responses, together with written work and/or manipulative activity, as observable outcomes of problem solving. The method used to analyze or score these "protocols" is of particular interest. Newell and Simon (1972) seek to model the problem solver as an information processing system, and therefore analyze protocols accordingly. This sort of analysis ultimately requires constructing a computer program that simulates the individual's problem solving, reproducing the information processing identified in the protocol (e.g., Anzai and Simon, 1979).

Schoenfeld (Note 2) proposes using "plausible approaches" as a criterion for scoring subjects' protocols. For each approach, the solver is scored on whether awareness of the approach is shown, whether the approach is pursued, and how much progress toward solution is made with its use.

A great deal of research has been devoted to developing scores reflecting the use of particular heuristic processes during problem solving. McClintock (1979) reviews process-sequence coding schemes developed by Kilpatrick (1967), Lucas (1972), Kantowski (1974), and Blake (1976), all of which are based to some extent on problem-solving heuristics as discussed by Polya (1957, 1962, 1965). The most detailed of such schemes is that of Lucas, Branca, Goldberg, Kantowski, Kellogg, and Smith (1979), in which more than 50 different symbols are used to represent process and outcome categories. More recently I worked with Campbell, Carpenter, Frank, Kulm,

Schaaf, Smith, and Talsma on regrouping these into a more manageable system suitable for recording the processes used by junior high school students (Kulm, Campbell, Frank, Talsma, and Smith, Note 3). These coding schemes are especially interesting because they are at once highly detailed and task-independent, that is, the same coding system can be used with widely varying problems. Kulm et al. report considerable progress toward achieving intercoder reliability in this type of coding. Examples of the categories used are: R (reads the problem), S (separates information), F (draws figure, table), V (introduces variable or other notation), D (uses deduction), A (uses algorithm), T (assigns trial value to variable), P (plans final or intermediate goal, changes goal), L (uses like situations or analogy), etc. Modification of process-sequence coding schemes is continuing (see also Lucas, 1980; Zalewski, 1980).

The preceding summary of outcome measures is intended to be representative of the categories explored in this paper, and by no means exhausts the list of techniques that have been used.

Discussion

The Definition of Problem Solving

Let us now look back at the methods for observing "problem solving," with an eye to defining it. In all cases a task is posed which can be performed successfully or unsuccessfully in a long or short period of time. Information is always processed by the subject. Errors sometimes occur during the performance of the task, alternative strategies may exist, and the subject may be able to describe in words (concurrently or retrospectively) the strategy adopted. The task may be represented by a structured set of symbol-configurations which are manipulated by the subject. The subject's behavior can also be classified as a sequence of processes including but not limited to the use of standard algorithms.

We can imagine several possible situations. *1.* The subject "knows the answer" or is already at the goal when the task is posed. Operationally, the outcome measures discussed earlier do not detect any steps, processes, or significant time lag between the posing of the task and the correct response. *2.* The subject does not "know the answer," but "possesses a correct procedure" for arriving at the answer (operationally, arrives through correct processes at the correct answer or goal), and furthermore "knows" (can correctly state) that he or she possesses the procedure, and furthermore is able to describe the procedure verbally before carrying it out. The procedure may be a standard algorithm taught as part of the mathematics curriculum, or it may be a non-routine procedure which the subject possesses by virtue of prior learning or problem-solving experience. *3.* Same as *2,* but the subject is unable to describe the procedure in advance of carrying it out. *4.* Same as *3,* but the subject "does not know for sure" (cannot state with certainty) that he or she possesses the procedure until after the problem has been attempted. 5. The subject does not possess a procedure for arriving at the answer (operationally, does not arrive through correct processes at the answer or goal until additional information or assistance is provided).

Some definitions seem to draw the line between case 4 and case 5. For example, "the individual . . . does not possess an immediate answer, procedure, or algorithm which solves the problem" (Zalewski, 1980, p. 119). I am not sure how the word "immediate" is intended to qualify the definition. If the subject possesses a procedure for arriving at a procedure, the whole is still a procedure possessed by the subject. In any case we must reject defining "problem solving" to exclude cases 1-4, because we would then be ruling out any successful use of correct procedures from the domain of problem solving.

Duncker's definition draws the line between case 3 and case 4, saying in effect, "A problem is a task for which the person does not know whether he or she possesses a procedure, until it has been attempted." I disagree with this choice for two reasons.

First, none of the problem-solving measures we have discussed place direct emphasis on measuring whether the subject *knows* in advance that he or she has a procedure for performing the task. To ascertain this directly would require a structured interview in which the first question asked was, "Do you already know how to do this task?" An answer of "yes" would not automatically mean that the task was not a "problem," because the subject might be mistaken. Thus it would be difficult to distinguish clearly between problems and non-problems. It is evident that the subject's advance state of knowledge concerning whether he or she has a procedure is an interesting "metacognition" which may, however, be rather difficult to measure. (Silver, Branca, and Adams, 1980). It is the inclusion of such metacognitive criterion in the definition of problem solving which, in my opinion, accounts for much of the mystique often mistakenly associated with problem-solving.

Secondly, it does not appear to me that the psychological processes involved in solving tasks for which we know we have procedures are necessarily very different from those involved in solving tasks for which we do not know we have procedures. Consider, for example, a mathematician asked to solve a pair of simultaneous linear equations in two unknowns, who tries a few values mentally (without knowing whether a solution can be found rapidly this way), considers the idea of using a substitution method, and settles on a linear combination of the two equations to provide the most efficient route to the solution. The measurement processes that we have discussed might detect the use of at least two strategies, consideration of a third, and the application of decision criteria among the strategies. All that is missing is the element of uncertainty on the subject's part as to whether he or she can find the answer, and this does not seem to me to be enough reason to prevent such high-level cognitive processes from being called "problem solving."

Some definitions draw a line between case 2 and case 3. For example,

Lester (1980) suggests, "A problem is . . . a task for which there is no readily accessible algorithm which determines completely the method of solution." If the key idea here is "readily accessible," and if we interpret this as "accessible to verbalization," then we wind up considering only cases 1 and 2 outside the domain of problem solving. If the key idea is "algorithm" then case 2 would be considered problem solving only if the procedure involved is not rigorously algorithmic. While these may be useful distinctions, they again require subtle measurements to establish them. My own preference, therefore, is to consider cases 2-5 all within the domain of problem solving for research purposes. This is consonant with the definition offered by Skinner, as well as that of Greeno (1980). Its chief advantage is that it is easily characterized by means of the outcome measures discussed earlier. A task is a problem when steps or processes are detected between the posing of the task and the answer. This does not preclude the distinction for research or instruction purposes between the rote use of standard algorithms and other more interesting problem solving processes.

The Comparability of Problem-solving Studies

To replicate a problem-solving study, it is essential to make the tasks as nearly identical to the originals as possible, and to adopt similar methods of scoring problem-solving performance. In order to compare the results of two studies, the tasks may be held constant while the scoring system is varied, or a standardized scoring method may be used while the tasks are varied. Earlier it was suggested that some standardization of tasks would be a good idea. I would like to argue still more forcefully here for some standardization of scoring procedures.

There are many reasons why researchers might wish to use different tasks for different experimental purposes, but there seem to be far fewer impediments to the adoption of parallel scoring methods. This is especially true since standardized scoring procedures can be used in addition to whatever specialized scores suit the needs of the experimenter. For example Kulm et al. (Note 3) have taken tapes of problem solvers recorded by many different researchers and re-coded them according to a single process-coding scheme. Thus it becomes possible within limits to compare findings across a variety of tasks, population, and treatment conditions. By standardizing the outcome measures, we establish at least a minimal answer to the question, "What is the body of data which problem-solving theory needs to explain?" Such an answer is a prerequisite to creating a comprehensive theory of problem solving.

Reference Notes

1. Lester, F. K. "The Relevance of Psychological Problem Solving Research for Research in Mathematical Problem Solving." Paper presented at the conference on Issues and Directions in Mathematical Problem Solving Research, Indiana University, Bloomington, May 29-30, 1981.
2. Schoenfeld, A. H. "Measures of Students' Problem-solving Performance and of Problem-solving Instruction." Unpublished manuscript, 1980 (available from A. H. Schoenfeld, Dept. of Mathematics, University of Rochester).
3. Kulm, G., Campbell, P. F., Frank, M., Talsma, G., and Smith, J. P. "Analysis and Synthesis of Mathematical Problem-solving Processes." Paper presented at the Annual Meeting of the National Council of Teachers of Mathematics, St. Louis, Mo. 1981.

References

Anzai, Y. and Simon, H. A. "The Theory of Learning by Doing." *Psychological Review* 86, (2), 1979, pp. 124-140.

Barnett, J. C. "Toward a Theory of Sequencing." Unpublished Doctoral Dissertation, Pennsylvania State University, 1974. Dissertation Abstracts International 1975, *36*, 99-100A. University Microfilms, No. 75-15787.

Barnett, J. C. "The Study of Syntax Variables." *In* G. A. Goldin and C. E. McClintock (eds), *Task Variables in Mathematical Problem Solving.* Columbus, OH: ERIC, 1979, pp. 23-68.

Bessot, A. and Comiti, C. "Une Etude sur L'Approche du Nombre par l'Eleve du Cours Preparatoire." *Educational Studies in Mathematics 9,* 1978, pp. 17-39.

Blake, R. N. "The Effect of Problem Context upon the Problem-Solving Processes Used by Field Dependent and Independent Students." Unpublished Doctoral Dissertation, University of British Columbia, 1976. Dissertation Abstracts International, 1977, 37A, 4191-4192.

Branca, N. and Kilpatrick, J. "The Consistency of Strategies in the Learning of Mathematical Structures." *Journal for Research in Mathematics Education, 3* 1972, pp. 132-140.

Bruner, J., Goodnow, J. J., and Austin, G. A. *A Study of Thinking.* New York: John Wiley & Sons Inc., 1956.

Caldwell, J. "The Effects of Abstract and Hypothetical Factors on Word Problem Difficulty in School Mathematics." Unpublished Doctoral Dissertation, University of Pennsylvania, 1977. Dissertation Abstracts International, 1978, *38A,* 4637. University Microfilms No. 77-30178.

Caldwell, J. and Goldin, G. A. "Variables Affecting Word Problem Difficulty in School Mathematics." *Journal for Research in Mathematics Education, 10* (5), 1979, pp. 323-336.

Clements, M. A. "Analyzing Children's Errors on Written Mathematical Tasks." *Educational Studies in Mathematics, 11,* 1980, pp. 1-21.

Davis, G. A. *Psychology of Problem Solving: Theory and Practice.* New

York: Basic Books Inc., Publishers, 1973.

Dienes, Z. P. and Jeeves, M. A. *Thinking in Structures.* London: Hutchinson Educational, 1965.

Dienes, Z. P. and Jeeves, M. A. *The Effects of Structural Relations on Transfer.* London: Hutchinson Educational, 1970.

Duncker, K. "On Problem Solving." *Psychological Monographs, 58* (5), 1945.

Durnin, J. H. "Assessing Behavior Potential: A Comparison of Three Methods." Unpublished Doctoral Dissertation, University of Pennsylvania, 1971. University Microfilms, No. 72-17343.

Goldin, G. A. "Structure Variables in Problem Solving." *In* G. A. Goldin and C. E. McClintock (eds), *Task Variables in Mathematical Problem Solving.* Columbus, OH: ERIC, 1979, pp. 103-169.

Goldin, G. A. and Caldwell, J. H. "Syntax, Content, and Context Variables Examined in a Research Study." *In* G. A. Goldin and C. E. McClintock (eds), *Task Variables in Mathematical Problem Solving.* Columbus, OH: ERIC, 1979, pp. 235-276.

Goldin, G. A. and Gramick, J. "Algorithmic Structure and Children's Error Patterns in Arithmetic." *In* R. Karplus (ed), *Proceedings of the Fourth International Conference for the Psychology of Mathematics Education.* Berkeley, CA: Lawrence Hall of Science, 1980, pp. 14-23.

Goldin, G. A. and McClintock, C. E. (eds) *Task Variables in Mathematical Problem Solving.* Columbus, OH: ERIC, 1979.

Gramick, J. "Should Algorithms Be Taught to Conform to Observed Behaviors?" Unpublished Doctoral Dissertation, University of Pennsylvania, 1975. University Microfilms No. 76-03168.

Greeno, J. G. "Hobbits and Orcs: Acquisition of a Sequential Concept." *Cognitive Psychology, 6,* 1974, pp. 270-292.

Greeno, J. G. "A Theory of Knowledge for Problem Solving." *In* D. T. Tuma and R. Reif (eds), *Problem Solving and Education: Issues in Teaching and Research.* Hillsdale, NJ: Lawrence Erlbaum Associates Inc., 1980.

Hershkowitz, R., Vinner, S., and Bruckheimer, M. "Some Cognitive Factors as Causes of Mistakes in the Addition of Fractions." *In* R. Karplus (ed), *Proceedings of the Fourth International Conference for the Psychology of Mathematics Education.* Berkeley, CA: Lawrence Hall of Science, 1980, pp. 24-31.

Jerman, M. and Mirman, S. "Linguistic and Computational Variables in Problem Solving in Elementary Mathematics." *Educational Studies in Mathematics, 5,* 1974, pp. 317-362.

Jerman, M. and Rees, R. "Predicting the Relative Difficulty of Verbal Arithmetic Problems." *Educational Studies in Mathematics, 4,* 1972, pp. 306-323.

Kantowski, E. L. "Processes Involved in Mathematical Problem Solving." Unpublished Doctoral Dissertation, University of Georgia, 1974. Dissertation Abstracts International, 1975, *36,* 2734A. University Microfilms No. 75-23764.

Kilpatrick, J. "Analyzing the Solution of Word Problems in Mathematics: An Exploratory Study." Unpublished Doctoral Dissertation, Stanford University, 1967.

Kulm, G. "The Classification of Problem-Solving Research Variables." In G. A. Goldin and C. E. McClintock (eds), *Task Variables in Mathematical Problem Solving*. Columbus, OH: ERIC, 1979, pp. 1-21.

Laughlin, P. R. "Selection Strategies in Concept Attainment as a Function of Number of Persons and Stimulus Display." *Journal of Experimental Psychology, 70*, 1965, pp. 323-327.

Lester, F. K. "Research on Mathematical Problem Solving." In R. J. Shumway (ed), *Research in Mathematics Education*. Reston, VA: National Council of Teachers of Mathematics, 1980, pp. 286-323.

Lucas, J. F. "An Exploratory Study in the Diagnostic Teaching of Elementary Calculus." Unpublished Doctoral Dissertation, University of Wisconsin-Madison, 1972. Dissertation Abstracts International 1972, *32*, 6825A. University Microfilms No. 72-15368.

Lucas, J. F. "An Exploratory Study on the Diagnostic Teaching of Heuristic Problem-Solving Strategies in Calculus." In J. G. Harvey and T. A. Romberg (eds), *Problem-Solving Studies in Mathematics*. Madison, WI: Wisconsin Research and Development Center for Individualized Schooling, 1980, pp. 67-91.

Lucas, J. F., Branca, N., Goldberg, D., Kantowski, M. G., Kellogg, H., and Smith, J. P. "A Process-sequence Coding System for Behavioral Analysis of Mathematical Problem Solving." In G. A. Goldin and C. E. McClintock (eds), *Task Variables in Mathematical Problem Solving*. Columbus, OH: ERIC, 1979, pp. 311-325.

Luger, G. F. "State-Space Representation of Problem-Solving Behavior." In G. A. Goldin and C. E. McClintock (eds), *Task Variables in Mathematical Problem Solving*. Columbus, OH: ERIC, 1979, pp. 311-325.

Luger, G. F. and Bauer, M. "Transfer Effects in Isomorphic Problem Situations." *Acta Psychologica, 42*, 1978, pp. 121-131.

McClintock, C. E. "Heuristic Processes as Task Variables." In G. A. Goldin and C. E. McClintock (eds), *Task Variables in Mathematical Problem Solving*. Columbus, OH: ERIC, 1979, pp. 171-234.

Newell, A. and Simon, H. A. *Human Problem Solving*. Englewood Cliffs, NJ: Prentice-Hall Inc., 1972.

Newman, M. A. "An Analysis of Sixth-grade Pupils' Errors on Written Mathematical Tasks." In M. A. Clements and J. Foyster (eds), *Research in Mathematics Education in Australia, 1977*, Volume 1. Melbourne, 1977, pp. 239-258.

Nilsson, N. J. *Problem Solving Methods in Artificial Intelligence*. New York: McGraw-Hill Inc., 1971.

Polya, G. *How to Solve It*. (Second edition) New York: Doubleday & Co. Inc., 1957.

Polya, G. *Mathematical Discovery*, Vos. 1 and 2. *On Understanding, Learn-

ing, and Teaching Problem Solving. New York: John Wiley & Sons Inc., 1962, 1965.

Radford, J. K. and Burton, A. *Thinking: Its Nature and Development.* London: Wiley, 1974.

Reed, S. K., Ernst, G. W., and Banerji, R. "The Role of Analogy in Transfer Between Similar Problem States." *Cognitive Psychology, 6,* 1974, pp. 436-450.

Silver, E. A., Branca, N. A., and Adams, V. M. "Metacognition: The Missing Link in Problem Solving?" *In* R. Karplus (ed), *Proceedings of the Fourth International Conference for the Psychology of Mathematics Education.* Berkeley, CA: Lawrence Hall of Science, 1980, 213-221.

Skinner, B. F. "An Operant Analysis of Problem Solving." *In* B. Kleinmuntz (ed), *Problem Solving: Research, Method and Theory.* New York: John Wiley & Sons Inc., 1966, pp. 225-257.

Suppes, P., Loftus, E. J., and Jerman, M. "Problem Solving on a Computer-based Teletype." *Educational Studies in Mathematics, 2,* 1969, pp. 1-15.

Thomas, J. C. "An Analysis of Behavior in the Hobbits-Orcs Problem." *Cognitive Psychology, 6,* 1974. pp. 257-269.

Waters, W. "The Use and Efficiency of a Scanning Strategy in Conjunctive Concept Attainment." Unpublished Doctoral Dissertation, University of Pennsylvania, 1979. Dissertation Abstracts International, 1979, *40,* 1331A. (a)

Waters, W. "Concept Acquisition Tasks." *In* G. A. Goldin and C. E. McClintock (eds), *Tasks Variables in Mathematical Problem Solving.* Columbus, OH: ERIC, 1979, pp. 277-296. (b)

Webb, N. "Content and Context Variables in Problem Tasks." *In* G. A. Goldin and C. E. McClintock (eds), *Task Variables in Mathematical Problem Solving.* Columbus, OH: ERIC, 1979, pp. 69-102.

Zalewski, D. L. "A Study of Problem-Solving Performance Measures." *In* J. G. Harvey and T. A. Romberg (eds), *Problem-Solving Studies in Mathematics.* Madison, WI: Wisconsin Research and Development Center for Individualized Schooling, 1980. pp. 119-141.

Some Issues in Problem-solving Research in Mathematics Education

Harold L. Schoen

This paper presents some personal observations, raises issues, and poses questions concerning problem-solving research in mathematics education, particularly as it has been affected by information processing (IP) psychology. The paper deals first with measurement issues, and second, discusses an application of the IP model to research in teaching problem solving. An assumption is that the goal of mathematics education research is not merely to extend the problem domain of the psychologists to mathematical problems, but to improve ultimately the quality of mathematical instruction. The paper was motivated by the fine summary of psychological problem-solving research by Lester (Note 1).

Measuring Problem Solving Abilities

The basic measurement issue is one of defining problem solving, that is, determining the nature of that which is to be measured. While the tasks and testing procedures should depend on a test's purpose, which in turn depends on a theory of problem solving, measurement which operationally defines problem solving is being done by individual researchers now, and the results are shaping theory as it evolves. Hence, it is difficult to separate a theory-based definition of problem solving from its measurement. Following are a number of questions which researchers should consider as they plan measurement procedures.

General Questions

What outcomes should be considered? Are response times, strategy counts, measures of elegance of solution, process sequences, etc., educationally useful outcome measures? What are "problem-solving processes" and does the answer depend on our definition of problem solving? What are "qualitative measures?" Are they valid or must we always resort to quantitative data?

Questions Concerning Mathematical Content

How should mathematical content interact with general thinking skills in a measurement instrument? If the math content is to be a minor component, are we attempting to measure, and hence raise by our teaching strategies, a sort of verbal reasoning IQ? Is it possible to improve on previous unsuccessful efforts to do this? If the math content is to be a major component, how should this be measured? Should the content factor be separated in our tests from general problem-solving ability? Can it be? Since the reliability and validity of a test are functions of the students being tested, at what populations should we aim? Is "non-routine" problem solving only for educationally elite students? Does it make sense to call problems "routine" if, say, less than one-fourth of the target population can solve them after intensive instruction?

Questions Concerning Interview-Based Testing

Evidently being in a think-aloud interview does not affect the number of problems solved by a subject; does it affect the processes he/she might use? What do process counts mean? Why count those particular processes and not others? Are they the "best" in some sense? If so, in what sense? If different problems and different solvers elicit very different processes, what do process counts pooled across problem types and across subjects mean? Researchers are sometimes careful about interrater reliability, but will few problems with a small "N" provide internally consistent data? Can and should data from interviews be verified in other ways, e. g. IP researchers write computer simulation programs to verify their theories?

Questions Concerning Paper and Pencil Testing

Should a multiple-count (process-product or partial credit) scoring system be used? If scores arrived at by a multiple-count system have a correlation of approximately 1.0 with a right-wrong approach (when correction is made for the unreliability of the two measures), what is the advantage of the multiple-count system? A case for the diagnostic value of a multiple-count approach is often made, but if partial scores are totaled across items, how can two students be differentiated when both score, say, 20 points on a 10-item test with a maximum of five points per item, but when one was awarded two points on each item and the other scored five points on each of just four items? If partial scores are separated (e.g., for approach, plan, etc.), how can the dependency of the partial scores for a particular item be interpreted since, say, an incorrect approach affects partial scores of subsequent processes relative to that item? Is the score in a multiple-count system a measure of a student's ability to plan approaches to problems, to do the mathematics required, to get the correct answer or some combination of these? Can and should these abilities be measured separately? Is each a function to the same degree of the student's knowledge of mathematical content, reasoning ability, computational facility, etc.? Should problem-solving tests be developed only for research purposes or with possible classroom uses in mind as well? What type of test is likely to gain the most widespread use in classrooms?

Information-processing Model Applied to Instruction

In this section, a case is made for applying an IP model to research in teaching problem solving. This is one way that mathematics educators could borrow from cognitive psychology.

The language and concepts developed by IP problem-solving researchers have a good deal of explanatory power for many types of tasks, including the issue of conducting research in the teaching of mathematical problem solving. The viewpoint presented here is similar to that of Shulman (1975) and Lanier (1977).

The Problem of Teaching Problem Solving

Each day of the school year, individual mathematics teachers are faced with a situation like this one:

P_i: To teach my students in my setting to better solve the mathematical problems in my curriculum.

Here P_i is the situation facing the th_i teacher in some ordered set of all mathematics teachers. A working assumption of this paper is that the broad goal of mathematical problem solving research is to help this teacher as he/she faces P_i. Traditionally, mathematics education researchers have tacitly assumed that there are important principles which generalize across $\mathbf{P} = \{P_i\}$; hence, we see studies comparing methods of teaching that are greatly concerned with representative samples of students, teachers, and classes, and with control of extraneous variables so as to enhance the generalizability of the finds. The IP model suggests another point of view.

To begin, it seems clear that by any definition of a problem including any of the ten in Lester's (Note 1) paper, each P_i is a problem. To illustrate this, the typical problem-solving activities engaged in by a master teacher attempting to solve P_i are described by using the four-step model of Polya and others. The term P-environment refers to all aspects of a teacher's situation as he/she attempts to solve P_i.

1. *Understanding the problem*
   Identify the relevant and irrelevant aspects of the P-environment, i.e., what variables will affect the solution of P_i. Determine the levels of these variables in the P-environment. These would include (but not be restricted to) student, task, teacher and school variables. From this P-environment the teacher constructs a problem space.
2. *Devising a plan*
   Examine the problem space for information with which to choose strategies. Consider the effect these strategies will have on concomitant variables. Here the teacher uses condition-action links.
3. *Carrying out the plan*
   Apply the strategy; i.e., teach the unit or lesson using the materials, methods, schedule, etc. devised in Step 2. Of course, many sub-

problems must be solved, sometimes in a rapid-fire manner, as the strategy is applied. Master teachers have algorithmic strategies for solving many of these subproblems such as spontaneous management, feedback, and questioning issues.

4. *Looking back*

Give a test, or check the "correctness" of the solution by some other means of evaluation. Also, try the strategy in other P-environments. Go back over the other steps and make adjustments as appropriate. All this helps the teacher to develop a schema for other P-environments, and, hence, improves the teacher's ability to solve a new P_i in the future.

The teacher behaves like the problem solver in many other ways. For example, "functional fixedness" would explain lack of imaginative uses of the classroom environment in the teaching plan. The role of understanding, as described by Lester's paper in this volume, seems to fit; for example, teachers are likely to be successful if they understand the effect of environmental, subject, and task variables, and if they recognize what to look for in the setting (condition) which will help them make a wise teaching-strategy move (action). "Chunking" would explain the seemingly natural, rapid-fire decisions master teachers make in the face of large amounts of data coming from several directions, while the novice or ineffectual teacher falters. As in problem solving, chunking here appears to result from a combination of experience (i.e. being faced with related problems many times before), knowledge or understanding, and ability or personality characteristics. The IP concepts of task environment and problem space seem to fit nicely here, too.

In short, it seems sensible to consider each teacher, in each setting, at each level, and at each time as facing a different, but related, problem when dealing with P_i. Furthermore, this explanation fits the empirical evidence better than one which assumes that a large number of general rules apply to many situations. For example, if this explanation is accurate one would expect to see, as is the case, many conflicting and equivocal findings in classroom-based studies which compare teaching methods, since situation specific variables, singly or in combination, would override the effect of the method variable. This also would explain partially the difficulties that teacher educators often experience with teachers turning off to what they perceive as "irrelevant theory" and the related criticism by some that research has little application to classroom practice. Many "general principles" gleaned from traditional research studies may, in fact, have little application for practice in a particular setting.

Implications for Research

If each P_i is considered to be a problem, then the IP model can help organize research designed to learn how to make teachers better at solving P_i. Some possible implications are the following:

1. Very few properties are likely to be invariant across task environments and problem solvers. Thus, the goal of research would be not to derive generalized rules which apply across many situations, but to learn what it means to be a good problem solver in P-environments and to devise ways to develop such problem solvers. Think-aloud interviewing of teachers before and after instructional sessions, in a way analogous to the problem-solving process studies with students along with detailed observation of the instruction, may provide useful information.

2. General heuristics such as means-ends analysis and solving sub-problems are certainly a part of the master teacher's repertoire. Strategies for teaching mathematical problem solving can be thought of as specific heuristics for a teacher in a P-environment. Development of strategies should continue and their effectiveness in particular settings should be verified. Emphasis should be placed on developing many effective strategies with an understanding of the conditions under which they are likely to be most effective.

3. Teachers should be equipped with an understanding of the important variables within the task environment, a set of decision rules which lead to optimal teaching-strategy moves, and the skill needed to make good use of these teaching strategies. Thus, it seems important for researchers to conduct studies which improve understanding of classroom environments, of what the crucial variables are, of what conditions should lead to what actions, and of what actions will maximize the chances of attaining the desired outcome. In this area, mathematics education researchers could profit from the research on teaching and classroom environments in the general educational literature.

4. There should be a good deal of emphasis on the careful "reporting" of levels of environmental variables, student variables, teacher variables, task variables, and outcome variables. An important goal should be to explain as completely as possible the task environment for P_i in the study. This would lead to a greater understanding of the relevant and irrelevant task environment variables, the condition-action relations, and the likely outcomes of certain actions in particular P-environments, thereby providing a basis for helping the problem solver/teacher to better understand the problem, P_i.

5. Concern for large, representative samples when research relative to P is conducted could sometimes be counter-productive. Such studies are designed to control or randomize the variables which may be most important for solving a particular P_i. Instead, careful and complete studies of single classrooms, or a small number of similar ones, using a variety of research methodologies would be more appropriate. The results would not necessarily generalize externally to any other classroom; rather, they would provide heuristic-like guidelines for teachers in similar classes. As results accumulate from different settings, the properties which are invariant across task environments

and problem solvers may begin to emerge. One likely candidate is the consistent finding that doing more problems leads to increased ability to solve similar problems.

6. It would make sense to continue to study in highly controlled settings how mathematical learning occurs, how problems are solved, or how certain teaching strategies affect the learner. Knowledge and understanding of this kind would be valuable to a teacher as he struggles to solve P_i, especially at the understanding stage. It would help also to guide the development of appropriate curriculum materials and tests, among other things. However, as Easley (1975) asserts, "Learning theorists in education have . . . been attempting to design functional systems which teachers could be taught to operate. Unfortunately, these designs of instructional systems have never taken into account the *conceptual systems* of the teachers and students. Putting it in Skinnerian theory, the full set of operants were never observed and, therefore, were not taken into account in the programming" (p. 36). The IP model suggests that the full range of operants in the real-world of the classroom be taken into account.

Summary

In this paper, a number of questions were raised concerning the measurement of problem-solving ability. It was also proposed that information processing provides a potentially powerful model for research in instruction. A theme of the paper has been that researchers would be well-advised to take into account the realities concerning students and teachers in classrooms as they plan their measurement and research.

Reference Note

1. Lester, F. K. "The Relevance of Psychological Problem-solving Research for Research in Mathematical Problem Solving." Paper presented at the conference on Issues and Directions in Mathematical Problem-Solving Research, Indiana University, Bloomington, May, 1981.

References

Easley, J. A. "Thoughts on Individualized Instruction in Mathematics." *In Schriftenreihe des Institut fur Didaktik der Mathematik der Universitat.* Bielefeld, West Germany, 1975.

Lanier, P. E. "Research on Teaching Mathematics: An Overview." *In Proceedings of the Research-on-teaching Mathematics Conference.* Michigan State University, East Lansing, Michigan, 1977.

Shulman, L. S. (Chairperson). "Teaching as Clinical Information Processing: Panel 6 Report." *In* N. L. Gage (Ed.), *NIE Conference on Studies in Teaching.* Washington, D.C.: National Institute of Education, 1975.

A Model for Elementary Teacher Training in Problem Solving

John F. LeBlanc

Of all the issues related to research in mathematical problem solving, the question of how to train teachers to teach problem solving is the one that I feel most urgently needs attention. While it is true that we have relatively few answers to problem-solving research questions, some decisions must be made about the nature and structure of teacher training in problem solving in order to give teachers guidelines for presenting problem-solving activities to children. I wish to propose then an instructional model for educating elementary teachers in problem solving before they begin in-service training.

What is a "Problem?"

Since various and diverse definitions of problem have been proposed, I must specify the kinds of problems considered in this paper. Some problem types, such as routine computational examples (exercises) are omitted. Among the various categories of problems included are the *process* problem and the standard *textbook* problem. An example of each follows:

Process Problem. Tom and Sue visited a farm and noticed there were chickens and pigs. Tom said, "There are 18 animals." After a moment of silence Sue said, "Yes, and there are 52 legs." I bet you can't tell me how many there are of each kind of animal!

Textbook Problem. Tom and Sue were planning a party and decided to buy 6 cartons of cola. If there are 8 bottles in each carton, how many bottles would they purchase?

The distinguishing characteristic of a process problem is that it can be solved using a variety of strategies or "processes." While the standard problems found in textbooks and standardized tests may not be as rich a source as process problems for exemplifying various strategies, most teachers see them as the basis for problem-solving instruction. To draw distinctions between process and standard problems, then, seems needless and diminishes the potential effectiveness of teachers teaching some generalized problem-solving processes.

A Model for Teacher Training

There are four phases in the proposed model for training teachers to teach problem solving.

Phase I: Solving Problems

Teachers should be given some problems to solve. Class interaction should focus on how a problem might be solved rather than on the actual solution. As many different methods of solution should be elicited as possible. These methods should be labeled. (Teachers are almost always surprised that "guess and test" is a legitimate method—and one among several which should be promoted and refined.) The instruction during this phase should model the behavior one might hope teachers would display.

The problems selected for presentation to the teachers should be carefully selected and capable of solution using a wide variety of strategies. Although process problems should be the principal type of problems used (since they often necessarily elicit nonstandard solution strategies), textbook word problems should also be presented for multiple solution strategies. Again, teachers are often surprised at their own ingenuity in discovering these alternative strategies. That it is not "illegal" to use a strategy other than formal mathematical operations to solve a problem is a surprise to them and they are astounded at being congratulated for using an unusual approach. Teachers (and consequently their students) need to be cautioned that to encourage the use of nonstandard solution strategies is not to demean formal mathematical processes. Rather, it is to recognize that formal mathematical skills must be placed in their proper perspective as problem-solving tools.

More often than not, teachers believe they are unable to solve certain problems because they have forgotten the "correct" or "applicable" formula or formal mathematical technique. With proper encouragement, they can gain confidence in solving problems using a variety of strategies and some, ironically, seem to gain a new appreciation for formal techniques as well. For example, when presented with a problem for which using simultaneous linear equations might well be an efficient strategy, many teachers can solve it without using such a formal method. As a result of the group solution exercises, the teachers rapidly grow more confident about using informal strategies. At the same time they become less apologetic for not knowing the "correct" way to solve the problem.

Phase II: Polya's Model

Polya's conceptual model (*understanding, planning, proceeding* and *evaluating*) for solving problems is presented at this point. The first and last steps are most important, the first in particular.

To help students understand a problem, teachers need to learn techniques other than advising them to "read it again" or "read it carefully." One such technique is that of asking questions that promote understanding. These questions can be divided into three groups:

- Questions about the factual information given in the problem.
- Questions about what is wanted in the problem.
- Questions about the conditions stated.

Learning to formulate good questions carefully is best done through practice. Teachers in training should be encouraged to ask these questions as they work on the problems in class. Later, when they present problems to their students, they should ask such questions and encourage their students to ask themselves similar questions. The ultimate objective in this exchange is to train the problem solvers to ask themselves questions that promote understanding.

Teachers should recognize that their role in teaching the *evaluating* step in the problem-solving process is also vital. Teachers can emphasize the importance of evaluation in two ways:

- By reviewing the solution processes.
- By extending the problem to create similar and related problem situations.

As teachers in training solve the problems presented to them, the trainer should model desired teaching practices behavior by reviewing the strategies used to solve the problem and listing them. Further, the trainer can extend the problem so that the strategies used in solving the main problem can be reused and reemphasized. Problems can be extended in one or more of the following ways:

- By modifying the variables.
- By changing the conditions.
- By altering the problem questions.
- By changing the context.

Again, practice is important and the teachers should be asked to write extensions of several types for many problems. Teachers should be cautioned to check that the suggested extension problems can be worked out using similar strategies as the main problem.

Phase III: Having Children Solve Problems

The next phase in the proposed training sequence is to give the teachers a set of problems to present to their own students (or, if they are not presently teaching, to friends, relatives, etc.). They should be encouraged to ask (and elicit) questions for understanding, to note and record solution processes used by the problem solvers, and to use a few extension problems. It is important to emphasize that spending time on one problem and demonstrating the various ways it can be solved is more important than getting answers to several problems.

Invariably the results of presenting such problems to their own students are remarkably successful. Teachers are almost always astounded at the cleverness and ingenuity of their students. They are also surprised that some of their "slower" students prove to be among the more agile problem solvers.

The "debriefing" period following these classroom experiences is a very significant part of the training sequence. Teachers who were at first apprehensive about teaching problem solving are usually encouraged and elated at this point by their own and others' experiences. Of course, not all teachers will have experienced such success and an analysis of successful and unsuccessful moves should be made for their benefit.

The role teachers play in the *planning* and *proceeding* steps is minimal—one of helping students to get started and one of encouragement. In helping students to "get started" or move away from dead-end paths, teachers can remind students of other strategies they might try or suggest that they look at specific pieces of information. The problem solver should be encouraged to do the work independently and to try hunches, draw pictures and diagrams, and otherwise to use "nonstandard" approaches.

Phase IV: Compiling a List of Problems

The final step in the proposed sequence is to ask the teachers to compile a list of problems (at least half of which are process problems) for future use. One might also have the teachers make up two or three questions for understanding and two or three extension questions for each problem. Time should be spent reviewing the compiled problems and questions, and at this time some refinements on problem types can be made. Differences among puzzle problems, problems with "trick" solutions, process problems, and standard problems can be made. Discussion can also refine the concept of questions for understanding and extension questions.

Finally, the sets of problems can be modified and edited by each contributing teacher, and then duplicated and distributed as resource packets for the teachers. Such compiled sets provide a bridge between the training sessions and classroom practice, and are more likely to be used than commercial sets since the teachers are familiar and feel more comfortable with them.

It has been hypothesized that only a problem solver can teach problem solving and that the best way to teach problem solving is to provide problem solving practice. But my experience and the research of two of my doctoral students (Proudfit, 1980; Putt, 1978) suggest that by helping teachers gain confidence in solving problems and analyzing the processes, by providing them with specific techniques for interacting with students before and after the problem solving activities, and by stocking them with a set of problems to present to their students, one is successfully preparing teachers to instruct in problem solving. Teachers report that their own attitudes toward problem solving and instruction have been changed by this training and that their students *can* and *do* solve problems and enjoy the process.

As with any educational innovation, this model raises as many questions as it answers. Questions suitable for further research include:

1. Do teachers' attitudes and teaching practices improve with respect to problem solving? If so, are the changes long-term changes?
2. Likewise, do students of these teachers show improved attitudes and performances and do any of these improvements have a long-term effect?

3. Does the model apply to both process problems and standard story problems?
4. Is the model useful and appropriate for secondary mathematics teacher education as well as elementary teacher education?
5. Assuming the effectiveness of the model, it remains to be demonstrated that it is more effective than others. To what extent and in what ways is this model more or less effective than other approaches to teacher training in problem solving?
6. Which phases of the model play the largest role in changing teachers' attitudes and teaching practices, assuming there is a change?

Teachers of mathematics have not had the luxury of waiting for research to provide answers to questions associated with teaching and learning. Teachers of teachers have likewise found it necessary to make decisions about instructional practices without benefit of unequivocal research results. The model described in this paper was developed along empirical but non-experimental lines. What is needed now is research on these and other questions to determine the degree of the model's effectiveness and its adaptability to various teacher-training situations.

J. F. LeBlanc

References

Proudfit, L. *The Examination of Problem-Solving Processes by Fifth-Grade Children and Its Effect on Problem-Solving Performance.* Unpublished doctoral dissertation, Indiana University, 1980.

Putt, I. J. *An Exploratory Investigation of Two Methods of Instruction in Mathematical Problem Solving at the Fifth-Grade Level.* Unpublished doctoral dissertation, Indiana University, 1978.

Applied Problem Solving: Priorities for Mathematics Education Research*

Richard Lesh and Margaret Akerstrom

In mathematics education, research and curriculum development projects dealing with problem solving have tended to focus on textbook-type word problems and on content-independent "Polya-style" strategies and heuristics. This article contends that focusing on these concepts was initially ill-conceived and has proved unproductive. We provide examples of the kinds of problems, problem-solving situations, and attendant processes, skills, and understandings that we believe should receive priority attention in mathematics education.

The chief justification for giving instructional attention to word problems is that they presumably help students develop problem-solving abilities that are useful in everyday situations—for employment, for informed citizenship, for applications in other subject matter areas, and even for recreation. However, we have produced a large amount of data refuting this claim in a current NSF-funded research project dealing with applied problem solving in middle-school mathematics, and in a companion project dealing with the role of various representational systems in the acquisition and use of rational number concepts. Purportedly realistic word problems often differ significantly from their real-world counterparts with respect to degree of difficulty, processes most often needed in solutions, and errors most frequently committed (Lesh, Landau, and Hamilton, Note 1). Furthermore, if one looks at everyday situations in which mathematics is used, it becomes obvious that: 1. Many of the most important problem types are not at all similar to textbook word problems (Bell, Note 2); and 2. The processes, skills, and understandings that are most important in real situations are seldom stressed by mathematics educators (e.g., Polya 1957). The examples in this paper illustrate several problem types and processes important to mathematics use in realistic everyday situations. Lesh (1981) gives a rationale for focusing on successful and unsuccessful problem-solving behaviors of *average ability students,* on problems that involve easy-to-identify *substantive mathematics concepts,* and on *realistic problem-solving situations*—i.e., those in which a variety of outside resources are available, including calculators, other students, resource books, and teacher/consultants who supply specific facts and information upon request.

*This material was prepared with the support of National Science Foundation Grant Nos. SED-7920591 and SED-8017771. Any opinions, findings, conclusions, or recommendations expressed herein are those of the authors and do not necessarily reflect the views of the National Science Foundation.

117

Most information about problem-solving processes has come from research involving older students, exceptionally bright students, individual students working in isolation (often in artificial laboratory situations), or from situations involving highly contrived word problems, proofs, or puzzles involving underlying ideas of questionable mathematical worth. Research has neglected elementary or middle-school children, average (or below average) ability students, problems involving easy-to-identify substantive mathematical concepts, and *applied* problem-solving processes. For this reason, the processes educators discuss often appear to be unavailable to younger, or less-gifted students, and certain of these processes, such as modeling, have been substantially neglected.

Northwestern's Applied Problem Solving Project

A major component of our Applied Problem Solving project consists of a 16-week problem-solving course, involving 23 seventh graders, in which:

- Two days per week are devoted to video- and audio-taped observations of the students as they work in groups of three on problems. This is the best "candid camera" situation we could devise in a classroom setting.
- One day per week in which the children work individually either on worksheet/tests focusing on particular stages in problem solving (with special attention to "non-answer giving" stages) or on individual problem-solving processes, skills, or understandings (with special attention to "modeling" processes, metacognitive understandings, and applied problem-solving skills).
- Two days per week in which individual children either are observed in problem-solving situations similar to those used in the group problem-solving sessions, or are involved in clinical interview sessions dealing with the problem-solving stages and processes featured in the worksheet/tests.

The following problems are examples of some of the types we are using in the Applied Problem Solving project. They illustrate several of the processes, skills, and understandings that average ability youngsters need in order to use mathematical ideas in everyday situations.

Example 1: Planning a Vacation

The students were given the following problem:

The Parker family of Evanston, which consists of Mr. and Mrs. Parker and their children, John (age 15) and Cathy (age 10), plans to spend one week on Paradise Island, Bahamas during winter vacation (December 5-December 13, 1981). Mr. and Mrs. Parker want to play golf and tennis and the children enjoy tennis and snorkeling. The Parkers have set aside $4,000 for this vacation.

Use the travel brochures, hotel advertisements, and airline information that we have given you to plan their vacation. They want the best vacation possible for $4,000.

This problem is typical of many real applications of mathematics because the mathematics is a means to an end rather than an end in itself. The goal here is to make a decision or a comparison, not to calculate an answer. The children we observed did not assemble different vacation packages and choose the most satisfactory. Instead, they first made arbitrary decisions about airlines, hotels, etc., and then carried out ad hoc calculations to see whether they had spent all the money. If comparisons finally were made among several plans, the plans tended to be considered as wholes; seldom did students attempt to fit together the best parts of several competing plans. There was little evidence of planning to allow for comparisons or for modifying or checking individual spending schemes.

Furthermore, the cost of each plan was usually calculated without recording individual items and amounts, resulting in omissions and duplications. Because most real problems are not solved in a single step, it is important to record the results of intermediate steps. However, sixth and seventh graders have had little experience with problems in which it is important to record, justify, or even remember accurately the procedures they use to arrive at answers. Similarly, they have had little experience with multiple-step problems or with problems in which more than one mathematical topic (e.g., addition, percents, measurement, money, time) is involved.

It is possible that the availability of a calculator accentuated the tendency not to record information. Students simply maintained a running total for all the expenses they considered, neglecting to note specific individual expenses. This method retained some exact amounts (to the penny) within an overall calculation that included very rough estimates of some expenses (e.g., daily meals). The students apparently were unaware of or unconcerned about the inconsistencies in these behaviors. It is clear that many skills associated with the use of calculators in realistic situations have not been taught, and are not well understood by these children.

In many real-world situations, problem formulation is critical. In the vacation problem, part of the difficulty was determining the meaning of the word "best." To do so, students had to assign weights or quantitative cost-benefit values to qualitative information, opinions, or alternatives. Again, most sixth and seventh graders are quite deficient in these skills. In addition, applied situations like the vacation problem are often characterized by too much and/or too little information. On one hand, there was an overwhelming amount of information (i.e., air fares, hotel rates, package deals, etc). On the other hand, students usually requested some information that was not initially available. The problem also requires students to generate some of their own data, and to decide what data to collect and how to collect it. This problem type is considerably different from typical textbook word problems containing too much information (usually one extra numerical bit which the solver must ignore) or too little information (wherein the solver is expected to conclude that the problem cannot be solved).

Example 2: Measuring The Speed Of A Toy Car

In the Applied Problem Solving project, we wanted some of our problems to resemble those typically found in physics or other science books. One, an inclined plane problem, had the additional advantage of having been the focus of several psychological studies (e.g., Larkin, McDermott, Simon, and Simon, 1980). Unfortunately, inclined plane problems are unlikely candidates for real-world situations because quantities like the magnitude of forces, acceleration constructs, and coefficients of friction, etc., are seldom "givens" in real situations. Furthermore, our studies show that we need not go beyond elementary measurement and arithmetic to create problems that are sufficiently difficult to confuse most average ability adolescents. Consider the following problem:

> A 6-inch "pinewood derby" car was rolled down a 14-foot ramp. The ramp was tilted so that the speed at the bottom was slightly faster than walking speed. The car was allowed to roll off of the bottom of the ramp and onto the floor. It usually stopped rolling approximately 15 feet from the end of the ramp.
>
> The children were given a yardstick and a stopwatch and were asked, "How many miles per hour is the car going when it crosses this line (i.e., a line on the floor at the bottom of the ramp)?"

Like many real problems, no data were given at the start and part of the problem, in fact, was to decide what data to collect. In contrast, numbers are given in the word problems in most mathematics and physics books. Thus one of the most common types of errors is for youngsters to pick one of their "number processing routines" to produce an answer, sensible or not.

In fact, while many middle school children are quite precocious in their ability to find the speed of the car in feet per second, few have ever been introduced to the "units analysis" techniques that would allow them to convert quickly and easily the feet-per-second figure to miles-per-hour. The arithmetic of unit labels does not necessarily accompany corresponding training concerning the arithmetic of numbers. Many children, for example, are unaware of the fact that "miles per hour" can be thought of as "miles divided by hours," that unit labels can be cancelled, or that there are well-defined rules for manipulating unit labels.

Unlike textbook word problems, which usually describe a series of quasi-realistic situations in which a single idea or topic can be used, real problems seldom begin with an idea and then look for "applications." Instead, they begin with a situation and then look for relevant ideas that seldom fit into neat disciplinary categories or can be expressed in formulae or equations. The problem may require the integration of ideas from several topic areas including the following: sequences of arithmetic procedures (addition, division), various number concepts (rational numbers, rates, proportions), several qualitatively distinct measurement systems (length, time), and intuitive ideas from geometry, physics, or other areas. Many real-world uses of mathematics also involve assigning numbers or measures to qualitative information. These mappings bring forth a number of substantive mathematical issues that are seldom addressed in schools, but which occur regularly in everyday situations. The problem below exemplifies the difficulties associated with quantifying qualitative information and combining qualitative and various kinds of quantitative information.

Example 3: Determining Sports Camp Groups

Imagine that you are a counselor at a sport camp specializing in track and field. In preliminary events held the first day, the 12-year-old boys performed as follows:

Name	50-yd. run	100-yd. run	high jump	long jump	swim level
Andy	7.9 sec.	14.6 sec.	4'2"	10'9"	Bass
Brian	9.4	16.5	3'4"	10'0"	Sunfish
Charles	7.0	13.5	4'4"	11'6"	Shark
Doug	7.8	14.4	4'7"	12'0"	Bass
Eric	6.8	12.6	4'8"	12'2" ·	Shark
Fred	8.0	14.5	4'1"	10'9"	Trout
Greg	8.1	15.0	4'3"	10'6"	Trout
Herb	7.2	13.8	4'3"	9'6"	Shark
Jon	7.7	14.0	4'4"	10'11"	Trout
Kevin	7.4	14.0	3'8"	11'2"	Bass
Larry	7.7	14.2	4'6"	11'0"	Bass
Mike	8.0	15.0	4'0"	10'2"	Trout

(Swim levels, from highest to lowest are: Shark, Bass, Trout, Sunfish)
Based on this information, and the attached comments from their school coaches, you are to assign these twelve campers to three groups.

Coaches' comments
Andy: Tremendous effort has helped Andy make some good gains this season, but his performance is not consistent.
Brian: Poor attendance at practice, an unwillingness to follow team rules, and goofing around during practice have surely contributed to Brian's deplorable performance.
Charles: Charles has consistently worked hard.
Doug: Doug has good endurance in most running events but tends to be sloppy about his performance in some throwing events.
Eric: Eric is a dedicated and motivated team member.
Fred: Fred's fair ability on most events does not offset a poor attitude toward practice.
Greg: Greg's immature behavior has interfered with his progress this year and has occasionally disrupted the whole group. He seems to do the least work required to stay on our track team.
Herb: Herb has shown tremendous improvement this year, but his previous training had been rather weak so he still has a lot to learn.
Jon: Jon has a lot of natural athletic ability but hasn't worked up to his potential because of missed practices and failure to pay attention to advice from the coaches.
Kevin: Kevin is a natural athlete who hasn't worked up to his potential. I think he needs more of a challenge, more competition, to make him work harder.

Larry: Larry is a rather shy boy who doesn't care to be in the spotlight. He just doesn't seem comfortable with heavy competition.
Mike: Although Mike's ability in jumping events is not outstanding, he has made good progress this year in sprinting.

The "sports camp" problem is similar to the "vacation" problem because it too involves both "too much" and "not enough" information, and both qualitative and quantitative information. It also allows a variety of solutions and solution paths, varying in complexity and sophistication. Some groups generally "worked forward" from the data toward a solution, assuming from the beginning that their initial goal was to agree on a single "formula" for selecting, weighting, and combining data. Other groups used a "forward working" process that was more stochastic. That is, they did not try for one all-inclusive formula, using instead a "most important" subset of the data to make an initial sorting of athletes into groupings. Then, based on the remaining data, they modified the groupings to compensate for "obvious" injustices. Still other groups of students used a predominantly "working backward" variation of the above stochastic process. They began with an intuitively-generated trial solution which was gradually refined by matching it with the data and individual cases.

All of the above solution procedures were more holistic and cyclic than the "working forward/working backward" labels might suggest. That is, in each of the three procedures an initial conceptualization of the problem served to filter, organize, and interpret the information in the problem situation. Then, "modifications" (i.e., interpretations) of the problem led to more refined conceptualizations, which led, in turn, to additional modifications of the problem.

In their initial approach, most groups focused almost exclusively on a subset of the relevant information. Some groups focused on the qualitative information (i.e., the coaches' comments), ignoring the quantitative information altogether. Other groups ignored the qualitative information, and sometimes, a substantial portion of the quantitative information as well. The difficulty of combining several different kinds of quantities (e.g., distances, times, comments), an important element in this problem, soon became apparent.

For example, several groups attempted to make decisions based on an "average" of the running scores and jumping scores. To calculate the "average," one group simply added the four numbers and then divided by four. That is, to get a single measure from the scores 7.9 seconds, 14.6 seconds, 4'2", and 10'9", they calculated 7.9 + 14.6 + 4.2 + 10.9, and then divided by 4. They did not find it troubling that adding time and distance measures might be a questionable procedure, or that low scores were good in the running events whereas high scores were good in the jumping events. Also, it did not trouble them to treat 4'2" as 4.2, without any unit labels; this system worked relatively well for 4'2" but was woefully inadequate for 4'11".

General APS Problem Characteristics

The preceding examples represent just a few of the important problem types we have identified by examining ways mathematics is used in everyday situations. In the Applied Problem Solving project, most of our problems are designed to require 10 to 45 minutes for solution. A variety of solutions and solution paths are usually available, so evaluating the efficiency, risks, and benefits of alternative solution paths is also important. Pilot work on the project showed that one characteristic of a good, everyday, applied problem solver is the ability, upon confronting a problem, to quickly and accurately assess problem difficulty, needed resources, and time required for an adequate solution. A 30-second solution attempt is quite different from a 5- or 30-minute attempt. The problems involve straightforward uses of easy-to-identify ideas from arithmetic, measurement, or intuitive geometry; no "tricks" are needed. Contexts include family finances (balancing a checkbook, starting a lawn mowing business, purchasing a family car), measurement applications (estimating distances using different kinds of maps, purchasing enough wallpaper to cover a given room), predicting trends in tables of data, and other situations that might occur in the lives of youngsters and their families.

Because most of our problems emerge from concrete situations or from situations that are familiar to students, reading the problem is not a source of difficulty. The task is to find solutions, not (as in most word problems) to correctly interpret the meaning of the problem situation.

Artificial restrictions on solution procedures can result in unrealistic problem-solving situations even when the problems themselves are meaningful, interesting, and in other respects, "real." People seldom work in isolation, relying on only the power of their own minds to solve problems. Instead, good problem solvers extend their own powers by using *all* available resources, such as those available to the students in this project. They also quickly focus their attention on "critical" solution stages, using outside resources to complete other stages.

For many of our problems, the goal is not to produce a mathematical "answer." Instead, the goals include discovering a process or procedure that produces a given result or that will produce answers to a class of specific problems, investigating the assumptions underlying quantitative claims, and making non-mathematical decisions, comparisons, or evaluations using mathematics as a tool.

Additionally, many of the problems were designed so that the critical solution stages would be "non-answer giving" stages. In most puzzles or problems used to study problem solving, the starting situation and the desired end point are both given. More realistic problems often occur as "ouches" rather than as well-defined questions with clearly specified goals (in which the "problem" is to find a set of legal moves to get from the "givens" to the "answer"). Problem formulation, including the quantification of qualitative information, is an important stage in many real problem-solving situations. In some cases, there is an overwhelming amount of information, all of which is relevant, and the main difficulty is to select and organize the information that is most useful in order to find an answer that is good enough. In other cases, not enough information is available, but a useable answer must be found anyway. Or, additional information may need to be identified or generated as part of a solution attempt—the information may not all be given at the start. All of these characteristics of real problems are related to the use of conceptual models as filters to select, organize, and interpret information from real situations.

Unlike a great deal of the best current problem-solving research in substantive content domains, the theoretical descriptions and mathematical models we use do not seem to fit the characterization of information-processing systems. Our explanations of problem solving and concept formation tend to be more organismic than mechanistic, with our theoretical constructs bearing closer resemblance to many of Piaget's ideas than to artificial-intelligence models (Lesh, 1980; Lesh, Landau, and Hamilton, Note 1; Saari, Note 3). While we *do* treat the learner as an adaptive system whose interpretation of problems is influenced by internal models as well as by external stimuli, we do *not* treat mathematics as information to be processed, nor do we treat mathematicians as processors. For us, the mathematician or mathematics student is a "situation interpreter and transformer," and mathematics furnishes the conceptual models for making interpretations and transformations.

The Project on Rational Numbers

The ability to solve word problems may be an important school skill, but it has not been a focus of the Applied Problem Solving project. Processes which we believe are important for the solution of word problems have been investigated in our Rational Number project research (Behr, Lesh, and Post, Note 4; Behr, Lesh, Post, and Silver, Note 5). However, even here, the processes found to be most important are quite different from those emphasized in most prior word problem research.

The following examples illustrate some of the differences between typical word problems and their real-world counterparts. Each problem was given in three distinct forms: as a "word" problem, as a "concrete" problem, and as a "real" problem.

The "Word" Problem

The following problem was typed on an index card and handed to the student. Materials such as paper, pencils, clay pies, Cuisenaire rods, and counters were available, but their use was neither encouraged nor discouraged.

Jim's family ordered two pizzas for supper, one with sausage and one with mushrooms. Jim ate ¼ of the mushroom pizza, and ⅕ of the sausage pizza. How much did he eat altogether?

The "Concrete" Problem

This problem was presented orally, using pre-cut parts of 6-inch clay "pizzas" on cardboard "plates."

"Pretend these (pointing) are pieces of pizza. (Present the one-half piece.)
"How much is this? (pointing to the one-half piece)
(Present the one-third piece.)
"And, how much is this? (pointing to the one-third piece)
"Now, how much is this (pointing to both pieces simultaneously) altogether?"

The "Real" Problem

The problem was presented orally, using pre-cut pieces of 6x8-inch clay "cakes" on cardboard plates.

"Pretend that these are pieces of cake. (Present the one-fourth piece.)
"Yesterday, I ate this (pointing) much cake. How much did I eat?
(Put the one-fourth piece into a covered box—since it was eaten yesterday—and present the one-third piece.)
"Today, I will eat this much (pointing). How much will I have eaten altogether?"

The problems were administered to 80 children (16 in each grade, fourth through eighth) individually during interviews that each included 11 such problems. Five of the problems had to do with addition of unit fractions, five with multiplication of unit fractions, and one was a fraction-related "find the area" problem. Detailed descriptions of all of the tasks, and the results of this study, are presented in Lesh, Landau, and Hamilton (Note 1). Only the general conclusions of the study will be mentioned here.

On the "word" problem, all of the children used pencil and paper calculations to try to answer the question. Only five fourth and fifth graders, and fewer than half of the sixth, seventh, and eighth graders, got the correct answer. By far the most common error was that of adding the numerators and then the denominators, obtaining ¼ + ⅕ = ²/₉.

After the testing session, the "word" problem was acted out using clay "pizzas" like the ones used in the "concrete" problem. When the ¼ and ⅕ pieces were placed, side by side, on a single plate, most of the children acknowledged that the result was not ²/₉. Nonetheless, they were unable to provide the correct answer, and more than half *still* stated that the answer ²/₉ was correct for the "word" problem. In explaining the discrepancy between that calculated answer and the (unknown) apparent "real" answer, several children explicitly said something to the effect of "That is a math problem, and this is a real problem—they aren't the same." This suggests that they didn't really expect their math answers to correspond to "real world" results.

On the "concrete" problem, slightly more than 50 percent of the students insisted on using a pencil and paper computation procedure, even though, on this particular problem, we discouraged them from doing so. Among the children who attempted to use a concrete procedure, most based their answer on a perceptual estimate. Popular, incorrect "concrete" procedures included putting the two pieces together on a single plate and estimating the answer, putting the two pieces together on a single plate and attempting to subdivide the "whole" into "equal-sized pieces" (this was the usual source of error) that fit the clay pieces, and keeping the pizzas on separate plates— often giving a perceptually-based estimate of the sum, or else adding the number of parts (i.e., 2) and the total number of pieces (i.e., 9) to obtain an answer of $2/9$.

On the "real" problem, approximately 50 percent of the students used a pencil and paper computation procedure. The "concrete" and "real" problems differed in that one of the pieces was hidden (i.e., in the covered box) in the "real" problem and different shapes were used. Consequently, some representation of the missing piece was required, and the type of representation system employed was directly related to the subsequent type of solution procedure. Approximately 30 percent of the students indicated the size of the missing piece using a hand gesture or a pencil mark on the plate to demarcate the size of the invisible piece. The resulting answer was based on a perceptual judgment, with the size of the missing piece often being distorted in order to yield a "nice" sum like $½$. Most of the remaining 20 percent of the children drew a diagram to represent the missing piece in the "real" problem. Diagrams, however, were rarely used in the "concrete" or the "word" problems.

Comments Regarding the Three Problem Types

Success in the solution of the "word" problem was highly correlated with the ability to draw accurate pictures of fractions, or to represent fractions accurately using concrete materials. That is, the children were better at addition-of-fractions word problems if they could relate the written symbols to pictures or concrete objects. Ironically, however, both the "concrete" problem and the "real" problem proved to be slightly more difficult than the corresponding "word" problem (i.e., fewer people got them correct), and drawing pizza or cake diagrams actually made the problems more difficult. The explanation of this rather surprising phenomenon (which also occurred in other problems involving the use of Cuisenaire rods, and eggs and egg cartons) seemed to be that, if the written symbols were meaningful (i.e., if they were able to be related to pictures and materials), then adding number symbols was easier than adding pictures.

It is one thing for a child to know how to illustrate fractions like $¼$ or $1/3$ using clay "pizzas" or diagrams, and quite another to be able to illustrate $¼ + 1/3$ using the materials or pictures. Concrete materials that are useful for illustrating fractions may not be useful for illustrating *addition* of fractions.

That is, the addition of fractions may be more meaningful if it is built on a strong concrete understanding of individual fractions, but this does not imply that learning to add Cuisenaire rods or folded paper will facilitate the child's understanding of addition of fractions.

Young learners do not work in a single representational mode throughout the solution of a problem. They may think about one part of the problem such as the numbers, in a concrete way, but may think about other parts of the problem using other representational systems (i.e., actions, spoken language rules, or written symbol procedures). It is only the mature student who is able to work through an entire complex problem using a single representational mode or model.

Not only do youngsters shift from one representational mode to another during different solution steps to a problem, but some problems elicit multiple modes inherently from the start. For example, in real addition situations that involve fractions, the two items to be added may not always be two written symbols, two spoken symbols, or two pizzas—they may be one pizza and one written symbol, or one pizza and one spoken word. In such problems, which occur quite often in real situations, it is difficult to represent both addends using a single representational system.

The Rational Number project was especially interested in the interactions between internal and external representations of problem situations. Frequently, when a child solves a problem, an internal "interpretation" of the problem influences the selection (or generation) of an external representation. This external representation may involve a picture, concrete materials, or written symbols, but often models only part of the problem. For example, if the problem involved addition of fractions, then the child's first drawing might represent the fractions without any attempt to represent the addition process. Or, the initial drawing might more closely resemble a photograph of the problem situation than a schematic diagram depicting the underlying mathematical relationships. In either case, however, the external representation typically allows the child to refine his or her internal representation (interpretation)—which may lead to the generation (or selection) of a more refined external representation, and, subsequently, to a solution. Thus, external representations play a number of important roles in the acquisition and use of rational number concepts, including, among others, reducing memory load or increasing storage capacity, coding information in a form that is more manipulable, and simplifying complex relationships.

Conclusions

Some of the following conclusions proceed primarily from the Applied Problem Solving project, and others are drawn from the data on the Rational Number project. First, for real situations in which mathematics is used to solve problems, the most important problem types are quite different from those typically represented in textbook word problems. Second, the processes, skills, and understandings that are most important in real world problems seldom have been emphasized in research on word problem solving. Third, the applied problem solving processes that we have identified as "most important" tend to be quite easy to teach to youngsters, and they do improve everyday problem solving capabilities. Fourth, as the results from the three problem types in the Rational Number project illustrate, word problems are different not only from everyday applied problems, but also from their own carefully matched "real world" counterparts. Fifth, the processes needed for the solution of word problems are seldom sufficient for the solution of these "real world" counterparts. Indeed, few of those processes are ever needed in real situations.

One major category of processes that we have identified in real world problem solving has to do with the use of various representational systems and translations from one representational system to another. Some of the other modeling processes we have found to be relevant are the following: introducing suitable notation, simplifying the situation to fit one's model, refining one's models to adequately reflect reality in a problem situation, and investigating the usefulness or quality of model-based predictions. An important characteristic of these processes is that most of them contribute to the meaningfulness, as well as, to the useability of basic mathematical ideas.

Because of our emphasis on the role that substantive mathematical understanding plays in problem solving, we reject the dichotomy between content-independent processes and process independent content. We do not believe that students first learn an idea, then learn to solve problems using the idea, and finally learn to solve applied problems. When good students use mathematics ideas to solve everyday problems, they are less likely to "look for a similar problem" (a content-independent heuristic) than to "interpret the given problem from several perspectives." They tend to work forward, beginning with a qualitative understanding of the whole situation, and gradually working toward a more quantitative, step-by-step approach in which specific data are substituted into computational equations. Good applied problem solvers use powerful *content-related* processes rather than the general (and weaker) *content-independent* heuristic techniques which are in fact more often characteristic of poor problem solvers or of problem solvers who are put into situations for which they lack the necessary substantive mathematical ideas.

We believe that there is a dynamic interaction between basic mathematical concepts and many of the most important applied problem-solving processes. Since content-independent heuristics have proven to be basically unteachable and of dubious value, there can be little question that priority attention must now be focused upon those content-dependent processes that seem to be not only imminently teachable, but also surfacing time after time as those processes of critical importance in the solution of real-world problems.

Reference Notes

1. Lesh, R., Landau, M., and Hamilton, E. "Conceptual Models in Applied Mathematical Problem Solving." *In* R. Lesh and M. Landau (Eds.), *Acquisition of Mathematics Concepts and Processes*. New York: Academic Press, in preparation.
2. Bell, M. *A Taxonomy of Problems Involving "Real World" Uses of Arithmetic and Number Ideas*. Paper presented at the 1981 Annual Meeting of the American Educational Research Association, Los Angeles, California, April 1981.
3. Saari, D. G. *Cognitive Development and the Dynamics of Adaptation: Accommodation*. Unpublished manuscript, 1978. (Available from D. G. Saari, Mathematics Department, Northwestern University, Evanston, Illinois, 60201.)
4. Behr, M., Lesh, R., and Post, T. *Rational Number Ideas and the Role of Representational Systems*. Paper presented at the 1981 Annual Meeting of the American Educational Research Association, Los Angeles, California, April 1981.
5. Behr, M., Lesh, R., Post, T., and Silver, E. "Rational Number Concepts." *In* R. Lesh and M. Landau (Eds.), *Acquisition of Mathematics Concepts and Processes*. New York: Academic Press, in preparation.

References

Larkin, J. H., McDermott, J., Simon, D., and Simon, H. A. *Models of Competence in Solving Physics Problems* (Technical Report), Carnegie-Mellon University, Physics Department, Pittsburgh, PA., 1979.

Lesh, R. "On Intuition." *In* R. Karplus (Ed.), *Proceedings of the Fourth International Congress on Mathematical Education*. Berkeley, CA: Lawrence Hall of Science, 1980.

Lesh, R. "Applied Mathematical Problem Solving. *In Educational Studies in Mathematics*, 1981, 12(2).

Polya, G. *How to Solve It*. Garden City, NY: Anchor Books, 1957.

Problem-solving Research:
A Concept Learning Perspective

Richard J. Shumway

The chapters of this volume represent thoughtful, rather formal discussions of many aspects of research on mathematical problem solving. I have been asked to write about problem-solving research from the perspective of concept learning and offer observations "from a distance." In contrast to the other chapters I will offer an informal, personal view of current research efforts on mathematical problem solving.

At a recent research conference a noted British scholar remarked that many Americans seem to insult each other as a way of showing a special friendship. My hope is such an approach will stimulate discussion. I offer the following remarks in such a spirit of good friendship.

Exploring Some Personal Biases

Problem-solving Researchers: The Isolationists

- "Problem Solving is an elite form of learning. It is really quite special."
- "We don't know what a problem solving is, but if you read our definitions you'll know what it isn't." (Corollary: It probably isn't related to what you are doing.)
- "We just invented clinical research. It's going to save the world. You'd better start using it."
- "Psychologists rarely study mathematical problems; how could their research have any value to us?"
- "You get 10 points for using the words 'problem solving.' If you can use the word 'metacognition' you get 100 points." (A nonexample of such behavior is that of Alan Hoffer. He fines you $75 if you use the words "problem solving." I hope he doesn't hear about "metacognition.")

It isn't quite fair, but let me pretend to quote concept learning researchers too.

Concept Learning Researchers: The Overgeneralizers

- "Since all problem solving culminates in concept and skill learning, problem solving is pretty interesting to us too."
- "There are basically three learning types: skill learning, concept learning, and problem solving. Skill learning is S-R learning, and problem solving is just hard concept learning."
- "Problem solving is simply the learning of new concepts and/or the

application of known concepts in new situations."
- "Read *Psychological Abstracts.*"

Now that you have been alerted to some of my biases, let me further elaborate on each of them.

What Is Mathematical Problem Solving?

Problem-solving researchers are stuck here. Problem solving must be defined. "Are you kidding?" you say. "Problem solving is a very difficult concept to define." I'll even settle for some examples and nonexamples. Can you sort human behavior into examples and nonexamples of problem-solving behavior? Do you "know one when you see it"? I would suggest a two-stage process for generating a definition. First, identify a broad range of examples and nonexamples of mathematical problem solving on which a significant number of researchers can agree. Then, perhaps, a careful study of these examples and nonexamples will allow one to identify the critical defining attributes of problem-solving behavior and will produce a working definition. (As you can see, the problem of definition is a concept learning task.) I believe it is counterproductive to mathematics education for researchers to continue to hedge on a definition of problem solving. The current collecting of problem types and choosing to narrow the universe to problem types of a special kind are good beginnings toward defining problem solving.

In many ways it is unfair to say that problem solving has not been defined. However, many "definitions" resort to negative descriptive characteristics. For example, some definitions require the problem solver to be in the state of not knowing what to do. Others attempt to define problem solving by identifying those behaviors which are not problem solving. If the universe is known, defining the complement of a set is adequate to deduce the set itself. In the case of human behavior, the universe of possible behaviors is so vast that it is difficult to use such an approach effectively. The complement approach to problem solving tends merely to isolate problem-solving researchers. In an effort to define problem solving carefully, one takes up the general strategy of excluding rather than accepting the work of others. Researchers who are most fussy and scholarly about their definition, find themselves among the most isolationist, if they are using an exclusion strategy for definition.

The better one is at excluding other learning types, the fewer colleagues and relevant research are available for support. Because of the exclusionary approach, there is a conceptual isolationism among problem-solving researchers which is slowing the growth of new knowledge.

Clinical Research Versus Experimental Research

The claim that all problem-solving researchers are overly infatuated with clinical research is surely a strawman. Scholars recognize the research approach is a function of the research goal. Certainly if one is interested in descriptions of present conditions or looking for correlational relationships, then the clinical research approach can be appropriate. But if one is interested in predicting future behavior and identifying cause-effect relationships, then the experimental research approach is appropriate. Both techniques require attention to questions such as sampling (size and nature), generalizability, reliability, validity, replicability, Type I and Type II errors, and scholarship. Perhaps we are defensive about clinical research because in the one-study-per-lifetime model of the researcher, the clinical part of a study is often the pilot study. Have we not matured to the point where we can appreciate the valuable contributions both clinical and experimental research bring to work on problem solving? The differences between the two are not in the quality of the technique or the learning models adopted, but rather in the essential difference between the goal of description and the goal of prediction. Few problems would not benefit from both approaches, and problem-solving research is no exception. We need to get on with the job and stop choosing up sides on the clinical-experimental pseudo-controversy of clinical versus experimental research. The "issue" is a non-issue.

Psychological Research: Relevant or Irrelevant?

Some mathematics educators dismiss psychological research as irrelevant to the problems in their field. The tasks used by psychologists are rarely mathematical so how can their generalizations be applied to research in mathematics education?

I have heard a similar tune in a different context. After seeing equivalence relations defined and used in "abstract" or "modern" algebra, undergraduates preparing to teach mathematics often declare there are no equivalence relations in school mathematics. It does not occur to them that counting, equals, congruence, similarity, and equivalence of fractions are meaningful examples of equivalence relations. Properties of equivalence relations derived in algebra are shared by all these examples. The purpose of the abstract algebra approach is to make the proofs once and then gain the broad application to all equivalence relations.

Perhaps the psychologist has simply abstracted the essential qualities of tasks to gain potential application to many disciplines. Is there a meaningful similarity betwen "red and square" and "four right angles and equal sides"? For example, Bourne and others have done a substantial amount of research on the logical operations: negation, conjunction, disjunction, conditional, and biconditional (Bourne and Dominowski, 1972). We are fortunate in mathematics to have defined clearly the significant role that logical operators play in all mathematical concepts and principles. Yet don't we claim, as the undergraduate, that these ideas from psychology are too abstract to have significant application to mathematics education?

Of course, we cannot argue for direct application of psychological results to teaching mathematics. Though we are talking about potentially isomorphic structures, we cannot be sure of the relevance mathematics has to the learning of the structure.

While the very existence of mathematics education as a discipline argues for mathematics making a difference, wouldn't 30 years of work on problem solving and concept learning by psychologists be of great help in understanding the difference? I think it would.

To fully exploit this body of work, it would help to learn the vocabulary of psychologists and use the same words for the same structures. Furthermore, just as finite group theory has been of help and interest to physicists, we should look at the abstract work of psychologists for potential applications to our work in mathematics education. Such a search might give us access to a tremendous number of new ideas.

In 1979, "mathematics educators" produced 554 research listings (Suydam and Weaver, 1980) and "psychologists" produced 13,684 research listings (Psychological Abstracts, 1980). We cannot afford to ignore such a large body of work on learning. In my view, most of the chapters in this volume make great strides toward appropriate use of psychological research.

Problem Solving: The "In" Word

There is little to say here that hasn't already been said. The heavy use of the words "mathematical problem solving" simply argues further for a need for a carefully-conceptualized definition of the phrase.

Concept and Skill Learning: Goals of Problem Solving

Why does someone engage in problem solving? What are the most direct, tangible benefits to the learner after engaging in solving a problem? Clearly one benefit is the problem solution. But, having just solved a problem, would it not be reasonable for one to attempt to consider the problem an example of a *class* of problems and attempt to identify a procedure or technique which could be used on any member of the class? That is, problem solving is time-consuming and hard work. It makes no sense to repeat the same problem solving again and again if one can identify a problem class and a procedure for solving problems of that class. In other words, problem solving often culminates in identifying a problem class *(concept learning)* and designing a procedure *(skill learning)* for solving problems in that class. In effect, the goal of problem solving often is to reduce the problem to learning types more easily handled—concept learning and/or skill learning. One could argue problem solving ends and concept learning begins when one begins looking back, identifying similar problems, and engaging in other post-solution activities.

Is Strategy Selection Concept Learning?

A key element in problem solving is the selection of strategies in solving a problem. What causes a person to choose a particular approach to solve a problem? Is drawing a picture always a good idea, or is there some basis for making such a choice? How does one decide to try counting strategies as a way to approach a problem? Are good problem solvers able to sort problems according to which strategies are likely to be successful? Is not strategy selection an example of concept learning?

Concept Learning: Is It Problem Solving?

One typical paradigm in concept-learning research is the presentation of a series of examples and nonexamples so that the subjects may deduce, by examining the examples, what concepts and/or attributes are being used to classify the examples. Often such a task requires the elimination of irrelevant attributes and the identification of a logical operation which is used with the relevant attributes. Examples from school mathematics might include fifth graders seeing word problems sorted according to whether addition, subtraction, multiplication, or division is the appropriate operation and attempting to see why; ninth graders seeing expressions sorted according to whether or not they can be factored as the difference of two squares and attempting to identify the critical features; calculus students seeing integrals sorted by whether or not they are solvable by integration by parts and attempting to identify the reasons. The examples being sorted must be new examples not sorted before. Is this problem solving? It seems that it is if the subjects doing the tasks find it difficult, and it isn't if they find it easy. Is problem solving, then, *difficult* concept learning?

The observation of Krutetskii (1976) and others regarding the important role relevant and irrelevant attributes play in problem solving suggest a natural link with concept-learning research. Are there others?

Advice to Problem-solving Researchers From a Concept Learning Researcher

- *Define Problem Solving.* There is much to be learned from identifying exactly what it is you are talking about.
- *Don't Be Elitist.* Look for linking ideas in the research of others even if they don't call it problem solving.
- *Read Psychological Research.* Just as topology is related to analysis and geometry, psychology is related to mathematics education.

Some Concept Learning References

The following list of references includes reviews and examples of recent concept learning research. They are provided as an aid to the reader who wants to begin looking at this body of research.

1. Sowder, L. Concept and principle learning. In Shumway, R. J. (Ed.), *Research in Mathematics Education.* Reston, VA: NCTM, 1980, 224-285. One mathematics educator's view of concept learning.

2. Bourne, L. E., Jr. and Dominowski, R. L. "Thinking." *Annual Review of Psychology,* Palo Alto, CA: *Annual Reviews,* 1972, *23,* 105-130. An excellent research review to 1972.
3. Erickson, J. R. and Jones, M. R. "Thinking." *Annual Review of Psychology,* Palo Alto, CA: *Annual Reviews,* 1978, *29,* 61-90. A continuation of Bourne and Dominowski's review to 1978.
4. Bourne, L. E. and Guy, D. E. "Learning Conceptual Rules II: The Role of Positive and Negative Instances." *Journal of Experimental Psychology,* 1968, *77,* 488-494. An example of the abstract psychologist's work, which has implications for mathematics education.
5. Tennyson, R. D., Woolley, F. R., and Merrill, M. D. "Exemplar and Nonexemplar Variables Which Produce Correct Concept Classification Behavior and Specified Classification Errors." *Journal of Educational Psychology,* 1972, 144-152. An example of the application of ideas from psychology to instruction; should have applications to mathematics education too.
6. Tennyson, R. D. and Park, O. "The Teaching of Concepts: a Review of Instructional Design Research Literature." *Review of Educational Research,* 1980, *50,* 55-70. Reviews educational research related to concept learning.
7. *Psychological Abstracts.* Abstracts of a large body of research in psychology. With practice, monthly issues can be scanned in 10 minutes, identifying from 5 to 10 relevant articles for your special interests.
8. Lester, F. K., Jr. "A Procedure for Studying the Cognitive Processes Used During Problem Solving." *Journal of Experimental Education,* 1980, *48,* 323-327. I caught one of our editors using concept learning to study problem solving.

Readings 2, 3, and 7 provide an overview of psychological research on concept learning. Readings 1 and 6 review concept-learning research and may be of value to problem-solving researchers. Readings 4, 5, and 8 are specific studies illustrating concept-learning research that has stimulated my thinking. This kind of research could provide ideas for applications to research in mathematical problem solving.

To illustrate this point, let me take the reading least likely to be perceived as relevant to mathematical problem solving and suggest some potential implications. Consider some of the results reported in the Bourne and Guy reference, reading 4.

"AI tasks were more difficult than RI tasks..." (Bourne and Guy, 1968, p. 490). Bourne and Guy found attribute identification tasks were more difficult than rule identification tasks. That is, a task was more difficult for subjects when they were told the rule and asked to identify the relevant attributes than when they were given the relevant attributes and asked to deduce the rule. Krutetskii (1976) found that contextual details (attributes?!) significantly influence mathematical problem-solving performance; that is many children have difficulty in dealing with contextual characteristics of problems. If

contextual characteristics of problems are regarded as relevant and irrelevant attributes, perhaps Bourne and Guy's finding helps explain Krutetskii's finding. When children are solving problems, the attributes may be much more difficult to deal with than the rules. We may need to give more attention to student responses to the contextual characteristics of a problem. Examining the work on these variables in concept-learning research may offer us further insight.

"Rules differed in difficulty with conjunction easiest and conditional most difficult..." (Bourne and Guy, 1968, p. 490). That is, "and" was easiest, "or" next, and "if-then" the most difficult. Other work in psychology including a fourth rule, biconditional or "if and only if," confirmed this ordering and added "if and only if" as the most difficult. Most problem solving involves conditional and biconditional arguments. Why are such rules more difficult? What variables interact with rule difficulty? Can we get some clues for mathematical problem solving from such information? Several significant interactions between variables and rule were identified and interpretations were offered (Bourne and Guy, 1968, p. 491-494).

"Mixed positive and negative instances . . . (were) associated with the fewest (trials) and all negatives with the most trials to solution." (Bourne and Guy, 1968, p. 490) It is clear that negative instances are an important variable. Their use with positive instances appears to be the most favored condition. The mixed sequences are most favored with conditional rules, those commonly encountered in problem solving. Perhaps negative instances would be valuable examples for students attempting to identify relevant attributes and rules in problem solving. I see at least three potential issues: *1.* What role do negative instances play in strategy selection? *2.* Does student generation of negative instances facilitate problem solving? (e.g., do negative instances facilitate student understanding of the conditional relationships within a problem?) *3.* What protocol techniques can be used to manipulate variables such as negative instances?

To further elaborate, consider the Tennyson, Woolley, and Merrill (1972) paper, reading 5. The independent variables were probability (difficulty), matching, and divergency. The dependent variables were correct classification, overgeneralization, undergeneralization, and misconception. The concept was trochaic meter. (Don't stop reading just because it isn't mathematics! The words overgeneralization and undergeneralization are supposed to keep you interested.) Two positive instances are *divergent* when the irrelevant attributes are as different as possible and *convergent* when the irrelevant attributes are similar. A positive and a negative instance are *matched* when the irrelevant attributes are as similar as possible.

Divergent positive instances, matched positive and negative instances, and mixed difficulty examples lead most effectively to correct classification. Whereas, divergency, non-matching, and difficult examples lead to overgeneralizations, divergency, matching, and easy examples lead to undergeneralizations, and convergency, nonmatching, and mixed

difficulty examples lead to misconceptions.
(Tennyson, Woolley, and Merrill, 1972).

Problem-solving researchers should be vitally concerned with ideas related to transfer, undergeneralization, and overgeneralization. What causes students to succeed or fail in generalizing strategies to new problem situations? What kinds of problems (examples) should students consider in an effort to generalize their solution or strategy to other related problems? Would understanding divergency and matching assist in identifying related problems? Could the work of Tennyson and others suggest methods, strategies, variables, and research questions which would be useful in problem-solving research?

The Bourne and Guy and Tennyson, Woolley, and Merrill references are rich in ideas for mathematical concept-learning research. However, they also appear to have rich implications and ideas for problem-solving research. Only a very small portion of the work on concept learning in psychology is related to the issue of negative instances. There are other areas of potential research focus such as class, probabilistic, or ordinal concepts, learning and utilization, reception-selection paradigms, frequency and hypothesis testing models, dimension saliency, and memory which are as carefully researched and perhaps even more relevant to research in mathematical problem solving. My current work in concept learning focuses on frequency of features of irrelevant attributes as a critical variable. One's focus in problem solving will surely lead to other variables and work in psychology not represented by the above examples. My examples are only given as an existence proof. The psychological literature will have to be studied in order to determine the replacement set for the existential quantifier relevant to a particular area of interest.

As evidenced by this volume, it seems we have a critical mass of researchers in mathematical problem solving interested in seriously examining what can be learned from the psychological literature. It is an exciting time. Nothing but productive new knowledge can be the result.

References

Bourne, L. E., Jr. and Dominowski, R. L. "Thinking." *Annual Review of Psychology,* Palo Alto, CA: *Annual Reviews, 1972, 23,* 105-130.

Bourne, L. E. and Guy, D. E. "Learning Conceptual Rules II: The Role of Positive and Negative Instances." *Journal of Experimental Psychology,* 1968, 77, 488-494.

Erickson, J. R. and Jones, M. R. "Thinking." *Annual Review of Psychology,* Palo Alto, CA: *Annual Reviews, 1978, 29,* 61-90.

Krutetskii, V. A. *The Psychology of Mathematical Abilities in Schoolchildren.* (J. Kilpatrick and I. Wirszup, Eds.). Chicago: University of Chicago Press, 1976.

Lester, F. K., Jr. "A Procedure for Studying the Cognitive Processes Used During Problem Solving." *Journal of Experimental Education, 1980, 48,* 323-327.

Psychological Abstracts. 1980, *64* (Nos. 1-12).

Sowder, L. "Concept and Principle Learning." *In* Shumway (ed.), *Research in Mathematics Education,* NCTM, 1980, 224-285.

Suydam, M. N. and Weaver, J. F. "Research on Mathematics Education Reported in 1979." *Journal for Research in Mathematics Education,* 1979, *11* (4), 241-315.

Tennyson, R. D. and Park, O. "The Teaching of Concepts: a Review of Instructional Design Research Literature." *Review of Educational Research,* 1980, *50,* 55-70.

Tennyson, R. D., Woolley, F. R., and Merrill, M. D. "Exemplar and Non-exemplar Variables Which Produce Correct Concept Classification Behavior and Specified Classification Errors." *Journal of Educational Psychology,* 1972, *63,* 144-152.